Advances in Thermal
Spray Technology

Advances in Thermal Spray Technology

Editor

Shrikant Joshi

MDPI • Basel • Beijing • Wuhan • Barcelona • Belgrade • Manchester • Tokyo • Cluj • Tianjin

Editor
Shrikant Joshi
University West
Sweden

Editorial Office
MDPI
St. Alban-Anlage 66
4052 Basel, Switzerland

This is a reprint of articles from the Special Issue published online in the open access journal *Materials* (ISSN 1996-1944) (available at: https://www.mdpi.com/journal/materials/special_issues/Thermal_Spray_Technology).

For citation purposes, cite each article independently as indicated on the article page online and as indicated below:

LastName, A.A.; LastName, B.B.; LastName, C.C. Article Title. *Journal Name* **Year**, *Article Number*, Page Range.

ISBN 978-3-03943-168-7 (Hbk)
ISBN 978-3-03943-169-4 (PDF)

© 2020 by the authors. Articles in this book are Open Access and distributed under the Creative Commons Attribution (CC BY) license, which allows users to download, copy and build upon published articles, as long as the author and publisher are properly credited, which ensures maximum dissemination and a wider impact of our publications.

The book as a whole is distributed by MDPI under the terms and conditions of the Creative Commons license CC BY-NC-ND.

Contents

About the Editor vii

Shrikant Joshi
Special Issue: Advances in Thermal Spray Technology
Reprinted from: *Materials* **2020**, *13*, 3521, doi:10.3390/ma13163521 1

Josef Daniel, Jan Grossman, Šárka Houdková and Martin Bystrianský
Impact Wear of the Protective Cr_3C_2-Based HVOF-Sprayed Coatings
Reprinted from: *Materials* **2020**, *13*, 2132, doi:10.3390/ma13092132 5

Bo Li, Yimin Gao, Cong Li, Hongjian Guo, Qiaoling Zheng, Yefei Li, Yunchuan Kang and Siyong Zhao
Tribocorrosion Properties of NiCrAlY Coating in Different Corrosive Environments
Reprinted from: *Materials* **2020**, *13*, 1864, doi:10.3390/ma13081864 17

Jan Medricky, Frantisek Lukac, Stefan Csaki, Sarka Houdkova, Maria Barbosa, Tomas Tesar, Jan Cizek, Radek Musalek, Ondrej Kovarik and Tomas Chraska
Improvement of Mechanical Properties of Plasma Sprayed Al_2O_3–ZrO_2–SiO_2 Amorphous Coatings by Surface Crystallization
Reprinted from: *Materials* **2019**, *12*, 3232, doi:10.3390/ma12193232 27

Heli Koivuluoto, Enni Hartikainen and Henna Niemelä-Anttonen
Thermally Sprayed Coatings: Novel Surface Engineering Strategy Towards Icephobic Solutions
Reprinted from: *Materials* **2020**, *13*, 1434, doi:10.3390/ma13061434 43

Bo Ram Kang, Ho Seok Kim, Phil Yong Oh, Jung Min Lee, Hyung Ik Lee and Seong Min Hong
Characteristics of ZrC Barrier Coating on SiC-Coated Carbon/Carbon Composite Developed by Thermal Spray Process
Reprinted from: *Materials* **2019**, *12*, 747, doi:10.3390/ma12050747 59

Pia Kutschmann, Thomas Lindner, Kristian Börner, Ulrich Reese and Thomas Lampke
Effect of Adjusted Gas Nitriding Parameters on Microstructure and Wear Resistance of HVOF-Sprayed AISI 316L Coatings
Reprinted from: *Materials* **2019**, *12*, 1760, doi:10.3390/ma12111760 75

Satyapal Mahade, Nicholas Curry, Stefan Björklund, Nicolaie Markocsan and Shrikant Joshi
Durability of Gadolinium Zirconate/YSZ Double-Layered Thermal Barrier Coatings under Different Thermal Cyclic Test Conditions
Reprinted from: *Materials* **2019**, *12*, 2238, doi:10.3390/ma12142238 85

Monika Michalak, Filofteia-Laura Toma, Leszek Latka, Pawel Sokolowski, Maria Barbosa and Andrzej Ambroziak
A Study on the Microstructural Characterization and Phase Compositions of Thermally Sprayed Al_2O_3-TiO_2 Coatings Obtained from Powders and Water-Based Suspensions
Reprinted from: *Materials* **2020**, *13*, 2638, doi:10.3390/ma13112638 99

Satyapal Mahade, Karthik Narayan, Sivakumar Govindarajan, Stefan Björklund, Nicholas Curry and Shrikant Joshi
Exploiting Suspension Plasma Spraying to Deposit Wear-Resistant Carbide Coatings
Reprinted from: *Materials* **2019**, *12*, 2344, doi:10.3390/ma12152344 121

Matthias Blum, Peter Krieg, Andreas Killinger, Rainer Gadow, Jan Luth and Fabian Trenkle
High Velocity Suspension Flame Spraying (HVSFS) of Metal Suspensions
Reprinted from: *Materials* **2020**, *13*, 621, doi:10.3390/ma13030621 . . . . **131**

Vasanth Gopal, Sneha Goel, Geetha Manivasagam and Shrikant Joshi
Performance of Hybrid Powder-Suspension Axial Plasma Sprayed Al_2O_3—YSZ Coatings in Bovine Serum Solution
Reprinted from: *Materials* **2019**, *12*, 1922, doi:10.3390/ma12121922 . . . . **149**

Daiki Ikeuchi, Alejandro Vargas-Uscategui, Xiaofeng Wu and Peter C. King
Neural Network Modelling of Track Profile in Cold Spray Additive Manufacturing
Reprinted from: *Materials* **2019**, *12*, 2827, doi:10.3390/ma12172827 . . . . **167**

About the Editor

Shrikant Joshi is a Professor in the Department of Engineering Science at University West, Sweden with nearly 30 years of experience in areas spanning Surface Engineering, Laser Materials Processing and now Additive Manufacturing. He is a Chemical Engineer by academic training, having obtained his M.S. and Ph.D. degrees from the Rensselaer Polytechnic Institute and University of Idaho, respectively, in USA. Prior to moving to Sweden, he had long stints as a scientist at a couple of premier, federally funded materials' research laboratories in India. Shrikant's past work has attempted to bridge basic research, technology development and its transfer for industrial implementation. His current areas of research are solution and solution-powder hybrid thermal spraying and additive manufacturing.

Editorial

Special Issue: Advances in Thermal Spray Technology

Shrikant Joshi

Department of Engineering Science, University West, 46132 Trollhättan, Sweden; shrikant.joshi@hv.se

Received: 6 August 2020; Accepted: 8 August 2020; Published: 10 August 2020

Coatings deposited utilizing different thermal spray variants have been widely used for diverse industrial applications. Today, various coating techniques belonging to the thermal spray family and spanning a vast cost–quality range have been embraced by the industry to either extend the longevity of components or enhance their performance, especially when these parts routinely operate in harsh conditions. The current state-of-the-art route for depositing the ceramic coatings is usually atmospheric plasma spraying (APS), while metallic and intermetallic coatings are sprayed by high-velocity oxy-fuel (HVOF) methods. Between them, the above spray methods and the vast portfolio of commercially available spray grade powders are capable of providing coatings that can combat premature surface degradation when industrial components are exposed to wear, corrosion, oxidation, high thermal load, etc. The immense versatility of the technique has already led to its numerous industrial uses, ranging from the advanced gas turbine requirements to the relatively more mundane needs of sectors such as textile, mining, pulp and paper and petrochemical sectors.

However, efforts continue to explore new potential applications to further expand this envelope. Two of the papers in this Special Issue focusing on impact wear [1] and tribocorrosion properties [2] of sprayed coatings, and another that seeks to augment mechanical properties via plasma spray deposition of multi-constituent amorphous coatings [3] are motivated by the above. Another paper explores novel surface designs to develop thermally sprayed icephobic coatings [4]. With the emergence of new engineering materials such as composites, there is also an interest in implementing thermal spray approaches for imparting them suitable protection, as exemplified by the contribution focusing on ZrC barrier coatings deposited on SiC-coated carbon/carbon composites by vacuum plasma spraying [5]. Post-treatment of thermal sprayed coatings by adopting approaches such as shot peening and laser remelting has also been a subject of considerable academic research. As a complement to such efforts, one of the papers deals with gas nitriding of HVOF-sprayed AISI 316L low-carbon austenitic stainless steel coatings [6].

Traditionally, thermal sprayed coatings have been realized employing powder feedstock, with the particle size typically being in the 10–100 μm range, with the lower end of this range being preferred for high melting point materials such as ceramics. Use of such feedstock, now commercially available for an exhaustive spectrum of material chemistries, results in splat sizes that are several tens of microns and consequently in coarse-structured coatings. However, there is growing interest in realizing fine-structured coatings using submicron and nanosized powders that can potentially yield enhanced functional performance. Such a feedstock injection methodology constitutes the basis for Suspension Plasma Spraying (SPS), which has been found capable of producing coatings with tailored microstructures, including the extremely porous to the very dense, vertically cracked, columnar, etc., and thus are not easily realizable when using a typical spray-grade powder feedstock. With the above approach providing a convenient pathway to deposit fine-structured coatings, thermal spraying with suspensions is perhaps the next frontier.

A vast majority of interest in SPS has hitherto been driven by the excitement of obtaining columnar thermal barrier coatings (TBCs). One of the papers in this Special Issue investigates SPS-derived double-layered Gadolinium Zirconate/Yttria-Stabilized Zirconia (YSZ) coatings [7]. The rapidly growing interest in this method is apparent from contributions that extend use of SPS to other materials such as oxides [8] and carbides [9], as well as to other high-velocity non-plasma spray processes [8,10].

It is also relevant to point out that the advent of axial injection capable plasma spray systems is a potential game-changer for use of liquid feedstock in the form of suspensions or solution precursors. This is by virtue of the fact that axial feeding enables far more intimate contact between the liquid feedstock and the plasma plume to facilitate thermal energy transfer and enable effective utilization of plasma energy. This advantage, which manifests in the form of higher throughputs, longer stand-off distances etc., has been harnessed in a couple of the above-mentioned studies [7,9]. The favourable thermal energy transport between the plasma plume and the suspension feedstock has also encouraged deployment of "hybrid" powder-suspension feedstocks to achieve unique coatings microstructures. One of the papers investigates the performance of such a hybrid powder-suspension sprayed Al_2O_3—YSZ coating in bovine serum solution [11].

Admittedly, it was not possible to include in this Special Issue other key areas that, too, continue to play a crucial role in the continued development of thermal spraying. These include, for example, evolution of new torch designs, advanced characterization of coatings, novel approaches for in-flight diagnostics and modelling of coating formation. The use of artificial intelligence/machine learning and data-driven modelling approaches, as illustrated in one of the papers [12], is also destined to play an important role in the future as thermal spray expands into new application domains such as additive manufacturing. Perhaps these can be the focus of a subsequent Special Issue.

References

1. Daniel, J.; Grossman, J.; Houdková, Š.; Bystrianský, M. Impact Wear of the Protective Cr_3C_2-Based HVOF-Sprayed Coatings. *Materials* **2020**, *13*, 2132. [CrossRef] [PubMed]
2. Li, B.; Gao, Y.; Li, C.; Guo, H.; Zheng, Q.; Li, Y.; Kang, Y.; Zhao, S. Tribocorrosion Properties of NiCrAlY Coating in Different Corrosive Environments. *Materials* **2020**, *13*, 1864. [CrossRef] [PubMed]
3. Medricky, J.; Lukac, F.; Csaki, S.; Houdkova, S.; Barbosa, M.; Tesar, T.; Cizek, J.; Musalek, R.; Kovarik, O.; Chraska, T. Improvement of Mechanical Properties of Plasma Sprayed Al_2O_3–ZrO_2–SiO_2 Amorphous Coatings by Surface Crystallization. *Materials* **2019**, *12*, 3232. [CrossRef] [PubMed]
4. Koivuluoto, H.; Hartikainen, E.; Niemelä-Anttonen, H. Thermally Sprayed Coatings: Novel Surface Engineering Strategy Towards Icephobic Solutions. *Materials* **2020**, *13*, 1434. [CrossRef] [PubMed]
5. Kang, B.R.; Kim, H.S.; Oh, P.Y.; Lee, J.M.; Lee, H.I.; Hong, S.M. Characteristics of ZrC Barrier Coating on SiC-Coated Carbon/Carbon Composite Developed by Thermal Spray Process. *Materials* **2019**, *12*, 747. [CrossRef] [PubMed]
6. Kutschmann, P.; Lindner, T.; Börner, K.; Reese, U.; Lampke, T. Effect of Adjusted Gas Nitriding Parameters on Microstructure and Wear Resistance of HVOF-Sprayed AISI 316L Coatings. *Materials* **2019**, *12*, 1760. [CrossRef] [PubMed]
7. Mahade, S.; Curry, N.; Björklund, S.; Markocsan, N.; Joshi, S. Durability of Gadolinium Zirconate/YSZ Double-Layered Thermal Barrier Coatings under Different Thermal Cyclic Test Conditions. *Materials* **2019**, *12*, 2238. [CrossRef] [PubMed]
8. Michalak, M.; Toma, F.-L.; Latka, L.; Sokolowski, P.; Barbosa, M.; Ambroziak, A. A Study on the Microstructural Characterization and Phase Compositions of Thermally Sprayed Al_2O_3-TiO_2 Coatings Obtained from Powders and Water-Based Suspensions. *Materials* **2020**, *13*, 2638. [CrossRef] [PubMed]
9. Mahade, S.; Narayan, K.; Govindarajan, S.; Björklund, S.; Curry, N.; Joshi, S. Exploiting Suspension Plasma Spraying to Deposit Wear-Resistant Carbide Coatings. *Materials* **2019**, *12*, 2344. [CrossRef] [PubMed]
10. Blum, M.; Krieg, P.; Killinger, A.; Gadow, R.; Luth, J.; Trenkle, F. High Velocity Suspension Flame Spraying (HVSFS) of Metal Suspensions. *Materials* **2020**, *13*, 621. [CrossRef] [PubMed]

11. Gopal, V.; Goel, S.; Manivasagam, G.; Joshi, S. Performance of Hybrid Powder-Suspension Axial Plasma Sprayed Al_2O_3—YSZ Coatings in Bovine Serum Solution. *Materials* **2019**, *12*, 1922. [CrossRef] [PubMed]
12. Ikeuchi, D.; Vargas-Uscategui, A.; Wu, X.; King, P.C. Neural Network Modelling of Track Profile in Cold Spray Additive Manufacturing. *Materials* **2019**, *12*, 2827. [CrossRef] [PubMed]

© 2020 by the author. Licensee MDPI, Basel, Switzerland. This article is an open access article distributed under the terms and conditions of the Creative Commons Attribution (CC BY) license (http://creativecommons.org/licenses/by/4.0/).

Article

Impact Wear of the Protective Cr_3C_2-Based HVOF-Sprayed Coatings

Josef Daniel [1,*], Jan Grossman [1], Šárka Houdková [2] and Martin Bystrianský [3]

[1] Institute of Scientific Instruments of the Czech Academy of Sciences, Královopolská 147, 612 64 Brno, Czech Republic; grossman@isibrno.cz
[2] Research and Testing Institute in Plzeň, Tylova 46, 301 00 Plzeň, Czech Republic; houdkova@vzuplzen.cz
[3] Regional Technological Institute, University of West Bohemia, 306 14 Plzeň, Czech Republic; mbyst@rti.zcu.cz
* Correspondence: jdaniel@isibrno.cz; Tel.: +420-541-514-339

Received: 2 April 2020; Accepted: 1 May 2020; Published: 4 May 2020

Abstract: High velocity oxygen-fuel (HVOF) prepared CrC-based hardmetal coatings are generally known for their superior wear, corrosion, and oxidation resistance. These properties make this coating attractive for application in industry. However, under some loading conditions and in aggressive environments, the most commonly used NiCr matrix is not sufficient. The study is focused on the evaluation of dynamic impact wear of the HVOF-sprayed Cr_3C_2-25%NiCr and Cr_3C_2-50%NiCrMoNb coatings. Both coatings were tested by an impact tester with a wide range of impact loads. The Wohler-like dependence was determined for both coatings' materials. It was shown that, due to the different microstructure and higher amount of tough matrix, the impact lifetime of the Cr_3C_2-50%NiCrMoNb coating was higher than the lifetime of the Cr_3C_2-25%NiCr coating. Differences in the behavior of the coatings were the most pronounced at high impact loads.

Keywords: HVOF; hardmetal; chromium carbide; dynamic impact test; impact wear; thermal spraying

1. Introduction

In many branches of industry, the surfaces of components are exposed to mechanical loading, the influence of an aggressive environment, high temperature, or even a combination of these features. To increase their lifetime, the surface of such components used to be coated by protective coatings. Among others, the technology of thermal spraying earns its position by offering a versatile solution for various kinds of applications and types of loading [1]. Of the required properties, the problem of wear resistance is most often addressed. If combined with high temperatures, the CrC-based hardmetal coatings, deposited by high-velocity oxygen fuel (HVOF) spraying technology, proves its superiority under harsh loading conditions [2–5].

The thermally sprayed chromium-based coatings were evaluated many times, in regards to the used deposition technology [6,7], the influence of deposition parameters [8], and morphology and phase composition of feedstock powder [9]. Their wear resistance under various loading conditions, including the high temperature, was analyzed [3,7,10].

Often, the Cr_3C_2-25%NiCr coatings are applied to protect the coated parts against erosion, particularly concerning the intended application in the power industry [11–16], where erosion by solid particle or water droplet particle is the most curtail for components lifetime. In these works, the mechanism of erosion wear was proposed, taking into account the influence of high temperature and oxidation [13]. It was found that in the aggressive environment, the preferential oxidation of CrC carbides takes place [2,14], leading to a degradation of the coating's microstructure and increased erosion wear [14]. To increase the resistance against corrosion and oxidation in aggressive environments, the application of CrC-based coatings with alternative matrix compositions was suggested [2,17].

The Cr_3C_2-50%NiCrMoNb coating increased corrosion and oxidation resistance due to the optimized matrix composition and comparable abrasive and sliding resistance and wear resistance [2,3].

Despite efforts, the mechanism of erosion wear is still not clearly explained due to the many factors taking part in the wear process. Other than the external conditions, the microstructure, phase composition, and state of internal stress are important. The materials with higher toughness can better accommodate the energy of impacting particles and suffer less brittle cracking [13]. In the case of hardmetals, the toughness of the material can be optimized by an increase of the soft metal matrix [18].

As surface dynamic loading appeared in various kinds of components, the need for its evaluation has led to the development of the unambiguous test. Knotek et al. developed a method of the dynamic impact test [19], originally focused mainly on the analysis of thin-film systems. During the dynamic impact test, the surface of the specimen is repeatedly impacted by ball indenter. Load force is well-defined and impact frequency is used to be a constant value. Using a dynamic impact test, it is possible to determine the impact wear or impact lifetime of the tested specimen [20–25].

The dynamic impact test was consequently adopted by the researchers focusing on the evaluation of thick thermally sprayed coatings [26–28], although the number of studies is rather limited. David et al. compared the HVOF sprayed coatings based on the numbers of impacts at various loads end evaluated the fracture modes [27]. Bobzin et al. analyzed the Cr_3C_2-NiCr coating impact wear by evaluation of impact craters created by ~ 10^5–10^6 impacts to obtain the coating failure modes and mechanisms [26]. In a recent study, Kiilakoski et al. evaluated the fatigue life of ceramic coatings exploiting low-energy impact conditions [28]. In this study, the impact wear was successfully related to coating cavitation wear resistance.

Although the first attempts to evaluate the impact wear resistance of several thermally sprayed coatings were completed, complex study of the impact wear of the Cr_3C_2-25%NiCr HVOF-sprayed coating including coating impact lifetime under various impact load remains absent. Additionally, the methodology of results evaluation needs to be established. The aim of this work is the complex study of impact wear and dynamic impact load limits of the HVOF-sprayed Cr_3C_2-25%NiCr and Cr_3C_2-50%NiCrMoNb coatings. Results obtained by impact testing were compared and discussed with respect to the microstructure of the coatings.

2. Experimental

2.1. Sample Preparation

The HVOF-sprayed coatings were deposited onto the flat surface samples of high-speed steel 5 mm thick cylindrical samples with a diameter of 20 mm. The coated surface was grit blasted before spraying to ensure sufficient adhesion of the coating to the substrate surface, using Al_2O_3; F20 abrasive media. Commercially available powders were used to spray the coatings—Amperit 588.074 was used for the preparation of the Cr_3C_2-25%NiCr coating and Amperit 595.074 was used for the preparation of the Cr_3C_2-50%NiCrMoNb coating. Both coatings were deposited using the HP/HVOF TAFA JP5000 spraying device (Praxair Surface Technologies, Indianapolis, IN, USA). The thicknesses of the sprayed coatings were set to 400 μm. The spraying parameters are summarized in Table 1.

Table 1. Coatings deposition parameters.

Material	Cr_3C_2-25%NiCr	Cr_3C_2-50%NiCrMoNb
Feedstock	Amperit 588.074	Amperit 595.074
Oxygen	823 L/min	872 L/min
Fuel	25.7 L/h	21.7 L/h
Barrel length	100 mm	150 cm
Spray distance	360 mm	330 mm
Traverse speed	250 mm/s	250 mm/s
Feed rate	70 g/min	76 g/min
Carrier gas	Nitrogen, 6 L/min	Nitrogen, 6 L/min
Offset	6 mm	6 mm

2.2. Dynamic Impact Test

Coatings behaviour under dynamic impact load was investigated using impact tester developed at the Institute of Scientific Instrumentation CAS in Brno, Czech Republic [20,21]. The tester is schematically described in Figure 1.

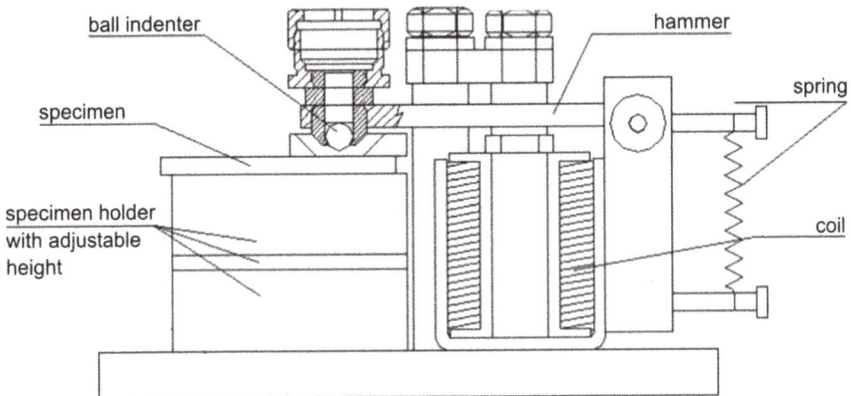

Figure 1. Dynamic impact tester developed at ISI CAS.

The impact hammer was driven electromagnetically; the amplitude of the loading force was regulated by an electric current in the solenoids. The ball-shaped tungsten carbide with a diameter of 5 mm was used as an impact indenter. The impacting frequency was set to 8 Hz. The indenter ball was adjusted to the unworn contact side before every test. Impact testing was carried out with the impact loads of 150 N, 200 N, 400 N and 600 N. Speed of the ball indenter before contact with specimen was in the range of 0.4 m·s^{-1} for the impact load of 150 N to 0.9 m·s^{-1} for the impact load of 600 N. Number of impacts were in the range from 1 up to 250,000. Every test was repeated three times for the elimination of surface inhomogeneity. All of the impact tests were carried out at room temperature.

The depth and radius of the impact craters were measured by the profilometer Talystep (Taylor Hobson, UK). The size and surface morphology of the impact craters were determined using confocal microscope Lext OLS 3100 (Olympus, Japan). Based on the work of Engel et al. for the physical vapor deposited (PVD) coatings [22], the methodology of critical crater volume determination was adapted and applied for evaluation of impact wear of HVOF sprayed coatings. In detail, the methodology is described in the next chapter.

3. Results and Discussion

3.1. CrC-Based Coatings Structure and Properties

Structure, phase composition, and mechanical and tribological properties of both of the studied HVOF-sprayed Cr_3C_2-25%NiCr and Cr_3C_2-50%NiCrMoNb coatings were described in detail by Houdkova et al. [2]. The Cr_3C_2-25%NiCr coating contained 25% of the Ni-Cr-based matrix, which surrounded carbide Cr_3C_2, Cr_7C_3, and $(Cr, Ni)_7C$ grains (Figure 2a).

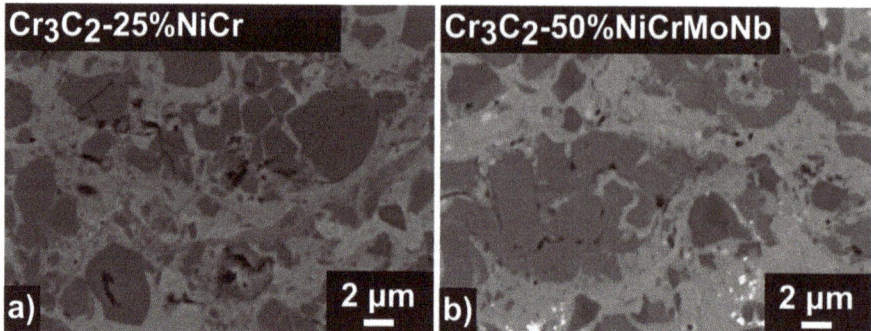

Figure 2. The microstructure of the studied coatings. Comparison of (**a**) the Cr_3C_2-25%NiCr coating and (**b**) the Cr_3C_2-50%NiCrMoNb coating.

Moreover, some amount of carbon was also incorporated into the matrix as a result of carbon dissolution during the spraying process. This various carbon amount in the matrix is represented as the shades of grey in the matrix in Figure 2a. On the other hand, the Cr_3C_2-50%NiCrMoNb coating contained 50% of the Ni–Cr–Mo–Nb-based matrix (Figure 2b). A matrix with a higher amount of dissolved carbon surrounded Cr_3C_2 and $(Mo, Ni, Cr)_7C_3$ grains and Nb–C precipitates. Furthermore, the Ni-based matrix also contained fcc precipitates with small crystallites and a bigger lattice parameter, the so-called γ-phase [2].

In [1], the mechanical properties and wear resistance of both coatings were compared. It was shown that despite the slightly lower hardness of Cr_3C_2-50%NiCrMoNb, caused by the higher amount of metallic matrix, its sliding wear resistance and coefficient of friction were comparable to the Cr_3C_2-25%NiCr coating.

3.2. Impact Behaviour of the CrC-Based Coatings

The basic result of impact testing is the loading curve-dependence of the impact crater volume on the number of impacts [20–22]. According to Engel et al., the loading curve can be divided into three zones, as is schematically drawn in Figure 3 [22].

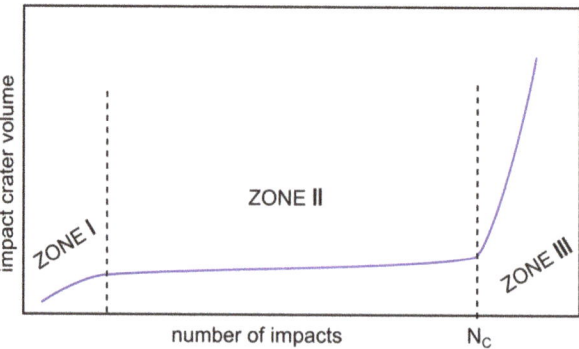

Figure 3. The scheme of the loading curve divided into three zones. Adopted from [22] and modified.

At the beginning of Zone I, the first impact causes deformation of the specimen. Any subsequent impact causes further deformation and the impact crater volume increases. However, the increment of the impact crater volume after the second impact is lower than increment after the first impact. As the number of impacts increases further, the increment of the impact crater volume decreases, and the system transits from Zone I to Zone II. This transition occurs at the low number of impacts and, for the investigation of the coating, impact lifetime is not important. In Zone II, the energy from the indentor is dissipated mainly in the form of inner stress in the specimen and just a minimal increase in crater volume is observed. In some materials, the formation of pile-ups around the impact crater can be observed. These pile-ups can be formed due to the material transport induced by impacting or due to the substrate deformation [29,30]. As the number of impacts increases further, the amount of stress increases up to a certain critical value. The number of impacts corresponding to this critical value, which is the critical number of impacts N_C, corresponds to the impact load limit. The critical number of impacts forms a boundary between Zone II and Zone III.

In Zone III, the impact crater volume rapidly increases with an increasing number of impacts, and coating tends to fail. The thin PVD coatings exhibit damage, delamination, and revealing of the substrate on the impact crater bottom and their N_C might be evaluated using both the rapid increase of the carter volume and the using the analyses of the coating delamination [11]. However, in the case of the thick HVOF coatings, no delamination or revealing of the substrate was observed. Thus, in this work, N_C was evaluated only using the rapid increase of the impact crater volume.

The impact wear was evaluated and compared for both CrC-based coatings at 150 N, 200 N, 400 N, and 600 N and analyzed. In Figure 4, the evolution of impact crater volumes (in mm^3) on the number of impacts are shown for the impact load 200 N.

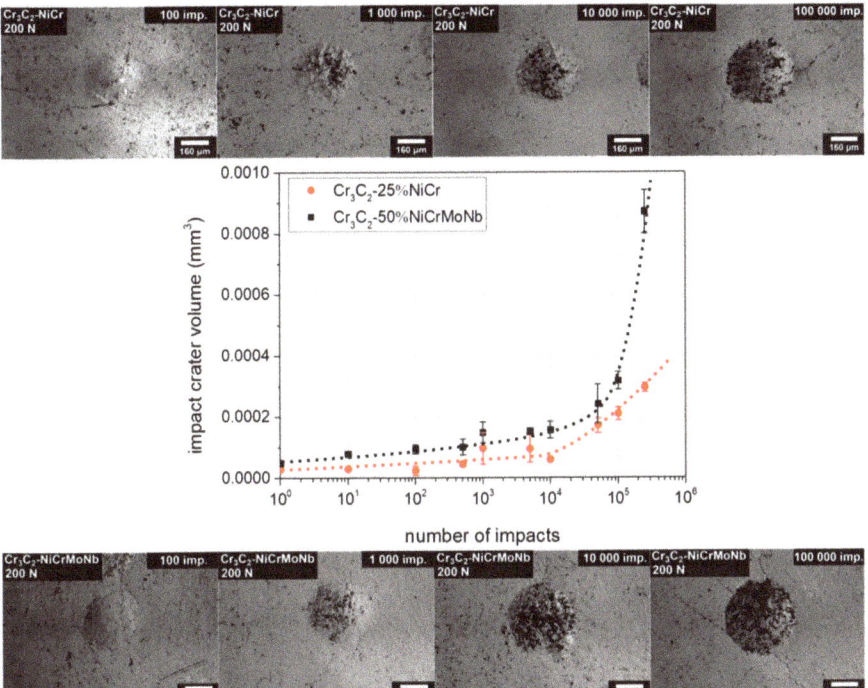

Figure 4. Comparison of the loading curves and impact craters of the studied coatings. The case for impact load of 200 N. Dotted lines were added as a visual guide.

Observed volumes of the impact craters of the Cr_3C_2-50%NiCrMoNb coating (black marks) were slightly higher than the volume of the Cr_3C_2-25%NiCr coating (red marks). Transition to the Zone III (rapid increase of the impact crater volume) occurred in the case of Cr_3C_2-25%NiCr coating earlier than in the case of the Cr_3C_2-50%NiCrMoNb coating. Thus, the critical number of impacts of the Cr_3C_2-25%NiCr coating was, for impact load 200 N, lower than the critical number of impacts of the Cr_3C_2-50%NiCrMoNb coating. Impact craters of the Cr_3C_2-25%NiCr and Cr_3C_2-50%NiCrMoNb coatings related to the 100, 1000, 10,000, and 100,000 impacts and an impact load of 200 N are also depicted in Figure 4. One can see that the dimension of the Cr_3C_2-50%NiCrMoNb impact craters was slightly higher than the dimension of Cr_3C_2-25%NiCr craters.

The loading curves of the Cr_3C_2-25%NiCr and Cr_3C_2-50%NiCrMoNb coatings obtained with a load of 600 N are depicted in Figure 5. Volumes of the impact craters of both coatings in Zones I and II were comparable. The rapid increase of the impact crater volume of the Cr_3C_2-25%NiCr coating (red marks) was observed at a lower number of impacts than for the Cr_3C_2-50%NiCrMoNb coating (black marks). Moreover, the impact craters of the Cr_3C_2-25%NiCr coating created by $\geq$ 10,000 impacts exhibited a large dispersion of the volume. Impact craters of the Cr_3C_2-25%NiCr and Cr_3C_2-50%NiCrMoNb coatings related to the 100, 1000, 10,000, and 100,000 impacts and a load of 600 N are also depicted in Figure 5.

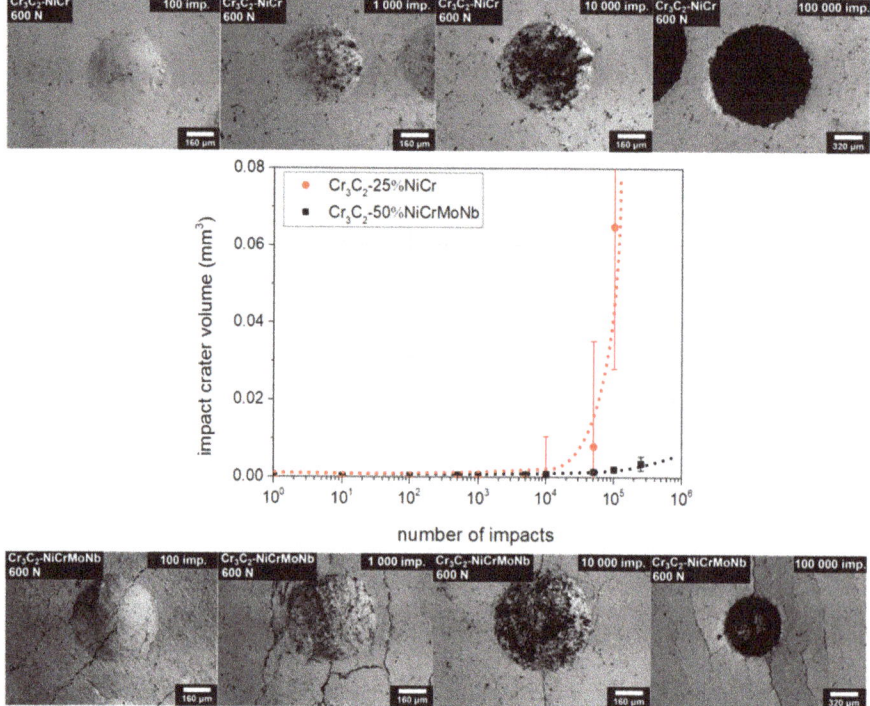

Figure 5. Comparison of the loading curves and impact craters of the studied coatings. The case for impact load of 600 N. Dotted lines were added as a visual guide.

Craters related to 100,000 impacts are depicted with twice the magnification of the others. The impact crater of the Cr_3C_2-25%NiC coating related to the 100,000 impacts was much larger than the corresponding impact crater of the Cr_3C_2-50%NiCrMoNb coating. Parallel microcracks through all micrographs in the case of the Cr_3C_2-50%NiCrMoNb coating are artefacts of mechanical damage of the coating.

Both the tested coatings exhibited radial cracking on the edge of the impact crater. The microcracks spread and expanded as the number of impacts increased. The dynamics of microcrack spreading can be explained using Figure 6.

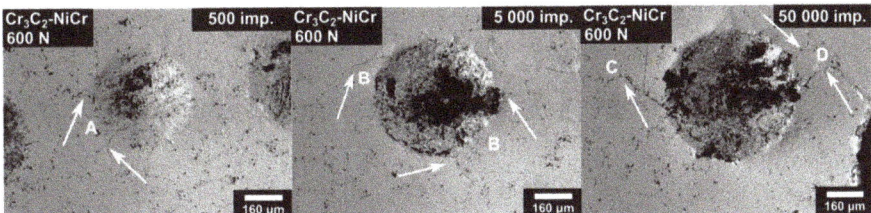

Figure 6. Dynamics of the microcracks spreading illustrated on the Cr$_3$C$_2$-25%NiC coating.

The impact craters of the Cr$_3$C$_2$-25%NiC coating related to 500, 5000, and 50,000 impacts are shown. At first, small radial microcracks were created on the edge of the impact crater (A). As the number of impacts increased, the number of radial microcracks grew and the size of the microcracks achieved the order of 100 μm (B). With further increase of impacts and thereby the amount of energy from the tester, the microcracks stopped to spread in the radial direction and tended to ramify (C). Some microcracks connected and created a closed area (D).

The first cracking, in the case of the dynamic load of 200 N, was observed in both coatings after 500 impacts. Both coatings exhibited similar microcrack size and density at the high number of impacts. However, in the case of the impact load of 600 N, the dynamics of microcrack spreading were different. The Cr$_3$C$_2$-25%NiCr coating exhibited the first indication of cracking after 10 impacts and after 50,000 impacts (see Figure 6) the 100–200 μm length microcracks ramified and connected themselves. On the other hand, the Cr$_3$C$_2$-50%NiCrMoNb coating exhibited the first indication of cracking after 500 impacts, and more than 300 μm length microcracks tended to create a connection at 250,000 impacts.

The process of evaluation of the critical number of impacts had to be slightly modified for the thick HVOF-sprayed coatings. The critical number of impacts, in the case of the thin PVD coatings, is estimated as the highest number of impacts before coating failure, like delamination or revealing of the substrate [19–22,30,31]. Such a critical number of impacts corresponds to the beginning of Zone III in the loading curve. In the case of HVOF-sprayed coatings, the critical number of impacts was estimated also as a beginning of Zone III in the loading curve. However, in the case of the observed large dispersion of the value of impact crater volume as shown in Figure 5, the critical number of impacts was estimated as the highest number of impacts before this large dispersion occurred. A similar methodology was used to evaluate the critical crater volumes for both coatings, tested at all of the impact loads.

Profiles of the impact craters of the Cr$_3$C$_2$-25%NiCr coating formed by the 100, 1000, 10,000, and 100,000 impacts are compared in Figure 7.

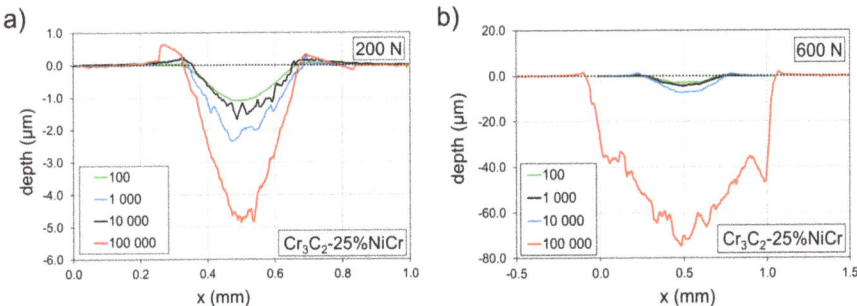

Figure 7. Profiles of the impact craters of the Cr$_3$C$_2$-25%NiCr coating. Comparison of the impact loads of (**a**) 200 N and (**b**) 600 N.

All profiles are related to craters depicted in Figures 4 and 5. Impact craters created by an impact load of 200 N (Figure 7a) and 100,000 impacts exhibited only small pile-ups on the edge of the impact craters. Since the depth of the impact crater was ~ 1% of the coating thickness, observed pile-ups were induced by the material transport rather than by the substrate deformation. Similarly, only small pile-ups were detected in the case of impact craters created by the load of 600 N (Figure 7b). The rapid increase of the impact crater after 100,000 impacts in Figure 7b corresponds to the rapid increase of the crater volume in Figure 5. The profile of this impact crater was considerably rougher in comparison to the other profiles.

Profiles of the impact craters of the Cr_3C_2-50%NiCrMoNb coating created by the 100, 1000, 10,000, and 100,000 impacts are shown in Figure 8.

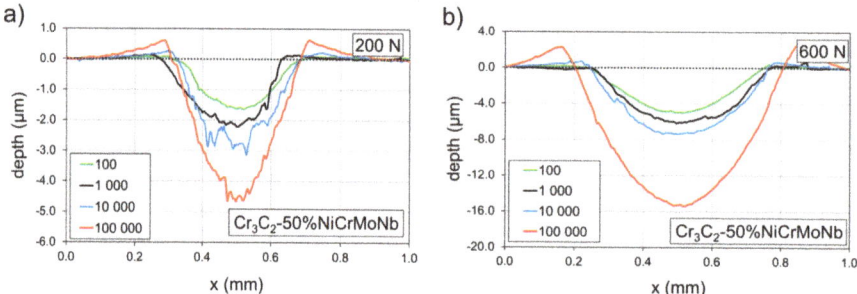

Figure 8. Profiles of the impact craters of the Cr_3C_2-50%NiCrMoNb coating. Comparison of the impact loads of (**a**) 200 N and (**b**) 600 N.

As in the previous case, all profiles are related to craters depicted in Figures 4 and 5. Craters formed under an impact load of 200 N (Figure 8a) and 10,000 and 100,000 impacts exhibited huge pile-ups on their edges. Similarly, craters prepared using an impact load of 600 N (Figure 8b) and by 10,000 and 100,000 impacts exhibited considerable pile-ups. Profile of the impact crater of the Cr_3C_2-50%NiCrMoNb coating prepared with a load of 600 N and 100,000 impacts was smooth in comparison with the corresponding impact crater of the Cr_3C_2-25%NiCr coating.

It was shown that the bonding of carbide Cr_3C_2 grains to the surrounding NiCr matrix in the Cr_3C_2-25%NiCr coating is a critical point with respect to the coating's wear resistance [4,5]. The formation of microcracks along the carbide grain boundaries and pulling-out of individual grains during wear tests were observed [2,7], leading to loss of coatings material. Therefore, we assume that the observed coarse profile of the crater prepared by the impact load of 600 N and 100,000 impacts (Figure 7b) in the Cr_3C_2-25%NiCr coating also resulted from the pulling-out of carbide grains. This can be explained as follows: when the number of impacts (or energy supplied from tester into the tested system) is sufficiently high, the stress in the material tends to form microcracks. Microcracks spread along the grain boundaries and cause the pulling-out of grains. Moreover, due to the random structure, microcracks also spread in random directions and pull-out a random amount of material. Thus, the volume of the crater was not the same for the same number of impacts and the average value of the volume of such impact craters exhibited large dispersion. This was observed in the case of the Cr_3C_2-25%NiCr coating and the load of 600 N (see Figure 5).

This claim is in good agreement with the dynamic of microcrack spreading observed at the surface of the coating and described above. The surface microcracks on the Cr_3C_2-25%NiCr coating prepared by the impact load of 600 N tended to connect at 50,000 impacts and create a closed area which can be easily pull-out. This microcrack spreading and thereby amount of this closed area is random and thus, the average value of the volume of such impact craters, exhibits a large dispersion.

A critical number of impacts of both coatings was determined for all the impact loads used—150 N, 200 N, 400 N, and 600 N. Dependence of the N_C on the impact load, the Wöhler-like curve, is depicted in Figure 9.

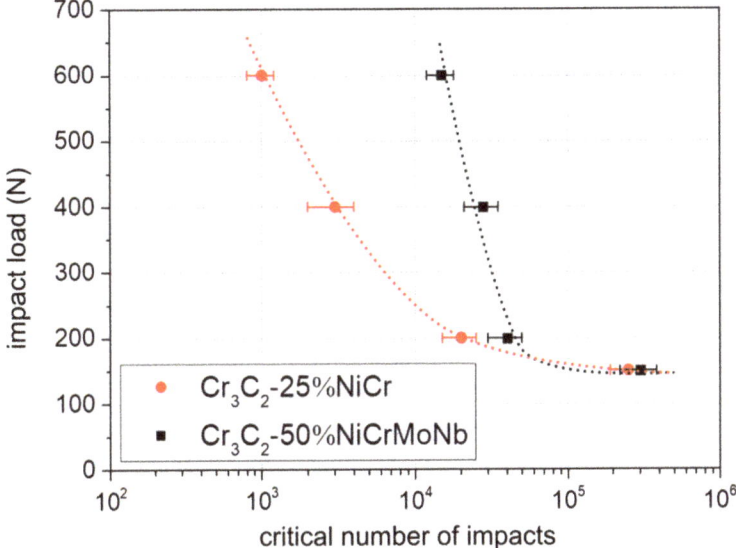

Figure 9. Dependence of the critical number of impacts on the used impact load. Comparison of the Cr_3C_2-25%NiCr and Cr_3C_2-50%NiCrMoNb coatings. Dotted lines were added as a visual guide.

The Cr_3C_2-25%NiCr coating (red marks) exhibited a lower impact load limit than the coating Cr_3C_2-50%NiCrMoNb (black marks). The difference in the impact load limit was the most distinct in the impact load of 600 N. The difference in the impact load limits became smaller as the impact load decreased. Finally, for the impact load of 150 N, the impact lifetime of both coatings was comparable.

A possible explanation of the different impact lifetimes is based on the different microstructure of the tested coatings. The coating Cr_3C_2-50%NiCrMoNb exhibited a higher number of soft matrices. Thus, energy supplied from tester to the coating was dissipated at first in the form of deformation and transport of material (pile-ups observed in Figure 8). The formation of microcracks required more energy, i.e., a higher number of impacts. On the other hand, the Cr_3C_2-25%NiCr coating exhibited a lower number of matrices and the possibility of material transport was limited. Supplied energy was rather dissipated into the formation of microcracks and the pulling-out of material. This resulted in a rapid increase in the impact crater volume at the lower number of impacts than in the case of the Cr_3C_2-50%NiCrMoNb coating. Thus, the Cr_3C_2-50%NiCrMoNb coating exhibited a higher critical number of impacts than the Cr_3C_2-25%NiCr coating.

The amount of energy supplied from tester to the tested system decreased with decreasing impact load. In the case of the lowest impact load of 150 N, the amount of supplied energy was insufficient to the formation of microcracks or even to the induction of material transport. Therefore, no differences in the impact load limit of both coatings were observed in case of the lowest impact load.

4. Conclusions

Impact wear of two HVOF-sprayed coatings—the Cr_3C_2-25%NiCr and the Cr_3C_2-50%NiCrMoNb—were investigated. The impact loads of 150 N, 200 N, 400 N, and 600 N were used. The results can be summarized as follows:

- The critical number of impacts of the thick HVOF-sprayed coatings was estimated using the loading curve and dispersion of values of the average volume of the impact crater.
- Observed dispersion of values of the average volume of the impact crater was the consequence of the different spread of microcracks in the coatings.
- The Cr_3C_2-50%NiCrMoNb coating exhibited, under the impact load of 200 N, a higher volume of impact craters, and the impact lifetime was nevertheless higher than for the Cr_3C_2-25%NiCr coating.
- The Cr_3C_2-50%NiCrMoNb coating exhibited higher impact lifetime than the Cr_3C_2-25%NiCr coating, probably due to the higher number of ductile metallic matrices.
- The difference between the impact lifetimes of the coatings was the most pronounced at high impact loads.

Author Contributions: Š.H. prepared tested coatings and analyzed their chemical and phase composition; M.B. analyzed coatings on SEM; J.D. and J.G. analyzed coatings using dynamic impact test; J.D. and Š.H. wrote the paper. All authors have read and agreed to the published version of the manuscript.

Funding: This research has been supported by the Czech Academy of Science (project L100651901) and by the institutional support for the long-time conception development of the research institution provided by the Ministry of Industry and Trade of the Czech Republic to Research and Testing Institute Plzeň.

Conflicts of Interest: The authors declare no conflict of interest.

References

1. Pawlowski, L. *The Science and Engineering of Thermal Spray Coatings*, 2nd ed.; John Wiley & Sons, Ltd.: Hoboken, NJ, USA, 2008.
2. Houdková, Š.; Česánek, Z.; Smazalová, E.; Lukac, F. The High-Temperature Wear and Oxidation Behavior of CrC-Based HVOF Coatings. *J. Therm. Spray Technol.* **2017**, *27*, 179–195. [CrossRef]
3. Matikainen, V.; Bolelli, G.; Koivuluoto, H.; Sassatelli, P.; Lusvarghi, L.; Vuoristo, P. Sliding wear behaviour of HVOF and HVAF sprayed Cr_3C_2-based coatings. *Wear* **2017**, *2017*, 57–71. [CrossRef]
4. Berger, L.-M. Application of hardmetals as thermal spray coatings. *Int. J. Refract. Met. Hard Mater.* **2015**, *49*, 350–364. [CrossRef]
5. Berger, L.-M. Hardmetals as thermal spray coatings. *Powder Met.* **2007**, *50*, 205–214. [CrossRef]
6. Hussainova, I.; Pirso, J.; Antonov, M.; Juhani, K.; Letunovits, S. Erosion and abrasion of chromium carbide based cermets produced by different methods. *Wear* **2007**, *263*, 905–911. [CrossRef]
7. Bolelli, G.; Berger, L.-M.; Börner, T.; Koivuluoto, H.; Matikainen, V.; Lusvarghi, L.; Lyphout, C.; Markocsan, N.; Nylén, P.; Sassatelli, P.; et al. Sliding and abrasive wear behaviour of HVOF- and HVAF-sprayed Cr3C2–NiCr hardmetal coatings. *Wear* **2016**, 32–50. [CrossRef]
8. Xie, M.; Lin, Y.; Ke, P.; Wang, S.; Zhang, S.; Zhen, Z.; Ge, L. Influence of Process Parameters on High Velocity Oxy-Fuel Sprayed Cr3C2-25%NiCr Coatings. *Coatings* **2017**, *7*, 98. [CrossRef]
9. Janka, L.; Norpoth, J.; Trache, R.; Thiele, S.; Berger, L.-M. HVOF- and HVAF-Sprayed Cr3C2-NiCr Coatings Deposited from Feedstock Powders of Spherical Morphology: Microstructure Formation and High-Stress Abrasive Wear Resistance Up to 800 °C. *J. Therm. Spray Technol.* **2017**, *26*, 1720–1731. [CrossRef]
10. Guilemany, J.M.; Miguel, J.; Vizcaíno, S.; Lorenzana, C.; Delgado, J.; Sanchez, J. Role of heat treatments in the improvement of the sliding wear properties of Cr3C2–NiCr coatings. *Surf. Coat. Technol.* **2002**, *157*, 207–213. [CrossRef]
11. Matthews, S.; James, B.; Hyland, M. Microstructural influence on erosion behaviour of thermal spray coatings. *Mater. Charact.* **2007**, *58*, 59–64. [CrossRef]
12. Matthews, S.; James, B.; Hyland, M. The role of microstructure in the mechanism of high velocity erosion of Cr3C2–NiCr thermal spray coatings: Part 1—As-sprayed coatings. *Surf. Coat. Technol.* **2009**, *203*, 1086–1093. [CrossRef]
13. Matthews, S.; James, B.; Hyland, M. High temperature erosion of Cr3C2-NiCr thermal spray coatings—The role of phase microstructure. *Surf. Coat. Technol.* **2009**, *203*, 1144–1153. [CrossRef]
14. Matthews, S.; James, B.; Hyland, M. High temperature erosion–oxidation of Cr3C2–NiCr thermal spray coatings under simulated turbine conditions. *Corros. Sci.* **2013**, *70*, 203–211. [CrossRef]

15. Tailor, S.; Vashishtha, N.; Modi, A.; Modi, S.C. Structural and mechanical properties of HVOF sprayed Cr3C2-25%NiCr coating and subsequent erosion wear resistance. *Mater. Res. Express* **2019**, *6*, 076435. [CrossRef]
16. Zhang, H.; Dong, X.; Chen, S. Solid particle erosion-wear behaviour of Cr3C2–NiCr coating on Ni-based superalloy. *Adv. Mech. Eng.* **2017**, *9*, 1–9. [CrossRef]
17. Fantozzi, D.; Matikainen, V.; Uusitalo, M.; Koivuluoto, H.; Vuoristo, P. Effect of Carbide Dissolution on Chlorine Induced High Temperature Corrosion of HVOF and HVAF Sprayed Cr3C2-NiCrMoNb Coatings. *J. Therm. Spray Technol.* **2017**, *27*, 220–231. [CrossRef]
18. Liu, J.; Bai, X.; Chen, T.; Yuan, C. Effects of Cobalt Content on the Microstructure, Mechanical Properties and Cavitation Erosion Resistance of HVOF Sprayed Coatings. *Coatings* **2019**, *9*, 534. [CrossRef]
19. Knotek, O.; Bosserhoff, B.; Schrey, A.; Leyendecker, T.; Lemmer, O.; Esser, S. A new technique for testing the impact load of thin films: The coating impact test. *Surf. Coat. Technol.* **1992**, *54*, 102–107. [CrossRef]
20. Sobota, J.; Grossman, J.; Buršíková, V.; Dupák, L.; Vyskočil, J. Evaluation of hardness, tribological behaviour and impact load of carbon-based hard composite coatings exposed to the influence of humidity. *Diam. Relat. Mater.* **2011**, *20*, 596–599. [CrossRef]
21. Daniel, J.; Souček, P.; Grossman, J.; Zábranský, L.; Bernátová, K.; Buršíková, V.; Fořt, T.; Vašina, P.; Sobota, J. Adhesion and dynamic impact wear of nanocomposite TiC-based coatings prepared by DCMS and HiPIMS. *Int. J. Refract. Met. Hard Mater.* **2020**, *86*, 105123. [CrossRef]
22. Engel, P.A.; Yang, Q. Impact wear of multiplated electrical contacts. *Wear* **1995**, *181*, 730–742. [CrossRef]
23. Bouzakis, K.-D.; Vidakis, N.; Leyendecker, T.; Erkens, G.; Wenke, R. Determination of the fatigue properties of multilayer PVD coatings on various substrates, based on the impact test and its FEM simulation. *Thin Solid Films* **1997**, *308*, 315–322. [CrossRef]
24. Heinke, W.; Leyland, A.; Matthews, A.; Berg, G.; Friedrich, C.; Broszeit, E. Evaluation of PVD nitride coatings, using impact, scratch and Rockwell-C adhesion tests. *Thin Solid Films* **1995**, *270*, 431–438. [CrossRef]
25. Bouzakis, K.-D.; Maliaris, G.; Makrimallakis, S. Strain rate effect on the fatigue failure of thin PVD coatings: An investigation by a novel impact tester with adjustable repetitive force. *Int. J. Fatigue* **2012**, *44*, 89–97. [CrossRef]
26. Bobzin, K.; Zhao, L.; Öte, M.; Königstein, T.; Steeger, M. Impact wear of an HVOF-sprayed Cr 3 C 2 -NiCr coating. *Int. J. Refract. Met. Hard Mater.* **2018**, *70*, 191–196. [CrossRef]
27. David, C.N.; Athanasiou, M.A.; Anthymidis, K.G.; Gotsis, P.K.; Neu, R.; Wallin, K.; Thompson, S.R.; Dean, S.W. Impact Fatigue Failure Investigation of HVOF Coatings. *J. ASTM Int.* **2008**, *5*, 101571. [CrossRef]
28. Kiilakoski, J.; Langlade, C.; Koivuluoto, H.; Vuoristo, P. Characterizing the micro-impact fatigue behavior of APS and HVOF-sprayed ceramic coatings. *Surf. Coat. Technol.* **2019**, *371*, 245–254. [CrossRef]
29. Batista, J.; Godoy, C.; Matthews, A.; Godoy, G.C. Impact testing of duplex and non-duplex (Ti,Al)N and Cr–N PVD coatings. *Surf. Coat. Technol.* **2003**, *163*, 353–361. [CrossRef]
30. Bantle, R.; Matthews, A. Investigation into the impact wear behaviour of ceramic coatings. *Surf. Coat. Technol.* **1995**, *74*, 857–868. [CrossRef]
31. Voevodin, A.; Bantle, R.; Matthews, A. Dynamic impact wear of TiCxNy and Ti-DLC composite coatings. *Wear* **1995**, *185*, 151–157. [CrossRef]

© 2020 by the authors. Licensee MDPI, Basel, Switzerland. This article is an open access article distributed under the terms and conditions of the Creative Commons Attribution (CC BY) license (http://creativecommons.org/licenses/by/4.0/).

Article

Tribocorrosion Properties of NiCrAlY Coating in Different Corrosive Environments

Bo Li [1,*], Yimin Gao [1], Cong Li [1,*], Hongjian Guo [2], Qiaoling Zheng [1], Yefei Li [1], Yunchuan Kang [1] and Siyong Zhao [3]

1. State Key Laboratory for Mechanical Behaviour of Materials, School of Materials Science and Engineering, Xi'an Jiaotong University, Xi'an 710049, China; ymgao@xjtu.edu.cn (Y.G.); zhengql@mail.xjtu.edu.cn (Q.Z.); yefeili@126.com (Y.L.); kangyc30@stu.xjtu.edu.cn (Y.K.)
2. School of Bailie Mechanical Engineering, Lanzhou City University, Lanzhou 730070, China; chinaghj2019@hotmail.com
3. Guangxi Great Wall Machineries, Hezhou 542800, China; wei9786@163.com
* Correspondence: libo616@mail.xjtu.edu.cn (B.L.); licong369@stu.xjtu.edu.cn (C.L.)

Received: 2 February 2020; Accepted: 3 April 2020; Published: 16 April 2020

Abstract: Atmospheric plasma spraying (APS) was taken to fabricate the NiCrAlY coating. The corrosion-wear properties of NiCrAlY coating was measured respectively under deionized water, artificial seawater, NaOH solution and HCl solution. Experimental results presented that the as-sprayed NiCrAlY coating consisted of Ni_3Al, nickel-based solid solution, NiAl and Y_2O_3. In deionized water, the coating with the lowest corrosion current density (i_{corr}) of 7.865×10^{-8} A/cm^2 was hard to erode. Meanwhile, it presented a lower friction coefficient and the lowest wear rate. In HCl solution, NiCrAlY coating gave the highest corrosion current density (i_{corr}) of 3.356×10^{-6} A/cm^2 and a higher wear rate of 6.36×10^{-6} mm^3/Nm. Meanwhile, the emergence of $Al(OH)_3$ on the coating surface could reduce the direct contact between the counter ball and sample effectively, which was conducive to the lowest friction coefficient of 0.24.

Keywords: corrosion-wear performance; dense structure; corrosion potential; corrosion rate; worn surface

1. Introduction

In engineering fields, wear often occurs under different corrosive circumstances leading to the degradation rate of engineering parts [1]. For instance, some mechanical parts utilized in the marine atmosphere, pulping and mining, suffer the collaborative destruction of corrosion and wear [2–6]. Meanwhile, the synergism of corrosion and wear decreases the service life of the material. In the process of friction, the passive film on the worn surface could be destroyed by friction force and the new passive film is hard to form, which would make the material suffering more serious damage. Normally, the corrosion-wear material loss is greater than the sum of corrosion and wear. Therefore, it is very imperative to improve the corrosion-wear resistance property of mechanical parts in different corrosive environments. To meet this requirement, the protective coatings are applied to protect the mechanical parts without changing the external structure. MCrAlY (M = Cobalt and/or Nickel) alloys with excellent oxidation resistance, corrosion resistance and wear resistance performance have been widely used in nuclear power, automotive and marine industries acting as the protective coatings [1,7–15]. J. Chen et al. investigated the tribocorrosion behavior of NiCoCrAlYTa coating in corrosion. The results showed that this kind of coating presented an extremely dense structural characteristic and excellent tribological performance in NaOH and HCl solutions [1]. M. Marcu et al. studied the microstructure and oxidation resistance of as-sprayed NiCrAlY/Al$_2$O$_3$ coating. The results presented that the as-sprayed NiCrAlY/Al$_2$O$_3$ coating has the best cyclic oxidation resistance with an oxidation rate of 2.62×10^{-12} $g^2 \cdot cm^{-4} \cdot s^{-1}$ at high temperature and good adhesion during the cyclic oxidization treatment [8]. Current researches mainly focus on the oxidation resistance, corrosion, mechanical and tribological performance of the coatings [16–22]. These materials are also used for reciprocating parts in corrosive environments [23], so the research of the wear-corrosion resistance is still important in the process of sliding. However, few researches pay attention to the synergy of corrosion and wear [24], and its mechanism is still unclear.

In this work, the tribocorrosion properties of NiCrAlY coating were studied and the synergistic mechanisms between wear and corrosion in different corrosive environments were discussed in detail. The objective of this paper is to research how corrosive environments affect the tribological behavior of NiCrAlY coating and the interaction degree between corrosion and wear. This research would provide usable direction to the NiCrAlY coating application in corrosive environments.

2. Materials and Methods

2.1. Coating Preparation

Gas atomized spherical $Ni_{22}Cr_{10}Al_{1.0}Y$ (wt.%) powder (53–106 μm) was bought from Sulzer Metco (Winterthur, Switzerland). The NiCrAlY coating was prepared by atmospheric plasma spraying (APS). The Inconel 718 alloy was sand-blasted, then ultrasonically cleaned with ethanol before spraying. The coating thickness was about 300 μm. The specific spraying parameters presented were: flow rate of Ar was 40 L/min; flow rate of H_2 was 5 L/min; spraying angle was 90°; feed rate of the powder was 42 g/min; voltage was 60 V; the current was 500 A and spray distance was 110 mm.

2.2. Characterization

The micromorphologies of cross-section and worn surface of this coating were measured by field emission scanning electron microscopy (FE-SEM, Tescan Mira 3, Bron, Kohoutovice, Czech Republic). A Philips X'Pert-MRD X-ray diffractometer (XRD; Cu-K_a radiation, current 150 mA, potential 40 kV, Philips, Eindhoven, The Netherlands) was utilized to analyzed phase composition. The phase compositions on the worn surface were analyzed by Czemy-Tumer Labram HR800 Raman spectrometer (Horiba, Paris, France).

2.3. Tribocorrosion Tests

The tribocorrosion experiments were tested in deionized water (pH = 7), artificial seawater (pH = 8.2), 0.1 M NaOH solution (pH = 13) and 0.1 M HCl solution (pH = 1), with reciprocating ball-on-disk tribometer (UMT, Karlsruhe, Germany). The schematic diagram is shown in Figure 1. The polytetrafluoroethylene (PTFE) does not corrode as it is chemically inert to corrosion. So it acted as the solution cell material. The Al_2O_3 ceramic ball acted as the counter ball, whose diameter was 5 mm. Before the friction experiment, the surface of the coating was burnished till the roughness close to 0.5 μm. The tests were performed at the conditions below: room temperature, 5 N normal load, 0.8 mm/s sliding speed, 3.5 mm amplitude and 60 min duration. Repeated experiments were tested in every corrosive environments. The color 3D laser scanning microscope (VK-9710, Keyence, Osaka, Japan) and SEM were utilized to analyze the worn surface. The wear rate was got by W = V/LF, where W represented the wear rate (mm^3/Nm), V represented the wear volume loss (mm^3), L represented the sliding distance (m) and F represented the load (N).

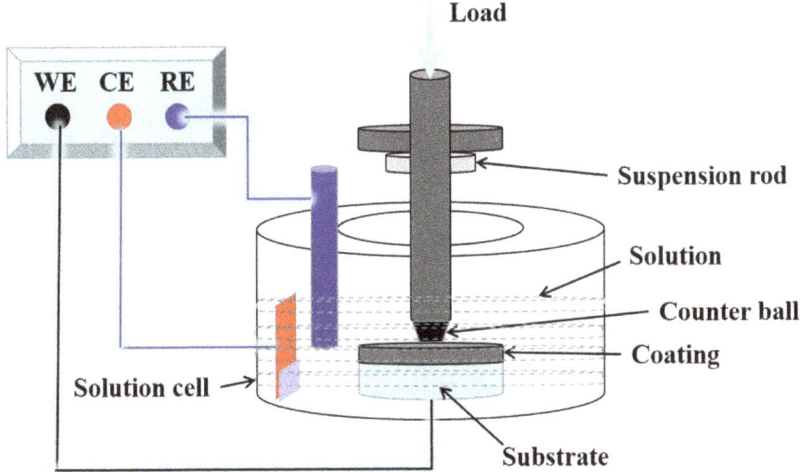

Figure 1. Reciprocating ball-on-disc tribometer schematic diagram.

3. Results and Discussions

3.1. Morphology and Composition of Powders and NiCrAlY Coating

Figure 2 presents the SEM micromorphology and XRD pattern of NiCrAlY powder. The spherical shape powder with a size of 53–106 μm (Figure 2a) exhibits satisfactory flowability and thus it is very beneficial to the feeding rate in the process of spraying [25]. The results of the XRD pattern show that the NiCrAlY powder composes of Ni_3Al, NiAl and nickel-based solid solution and has high crystallinity (Figure 2b).

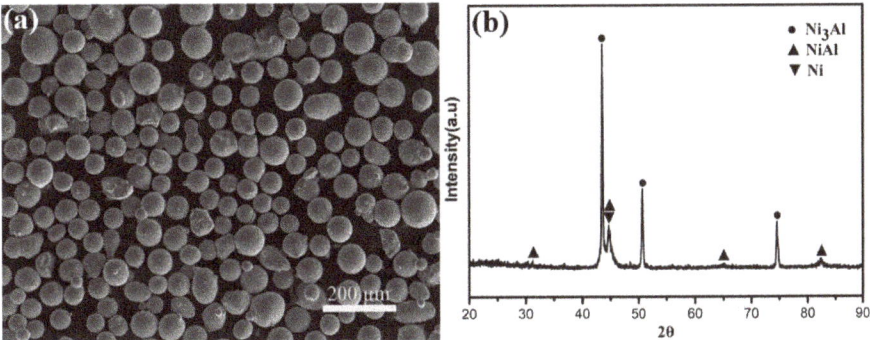

Figure 2. SEM micromorphology (a) and XRD pattern (b) of NiCrAlY powder.

Figure 3 presents the SEM morphology of the cross-section and diffraction pattern of NiCrAlY coating. The coating contains some cracks and pores. Meanwhile, every phase combines well and between any two phases have no evident cracks (Figure 3a). Compared with the NiCrAlY powder (Figure 3b), a new phase of Y_2O_3 formed on the coating, which could obviously increase the microhardness and strength [26].

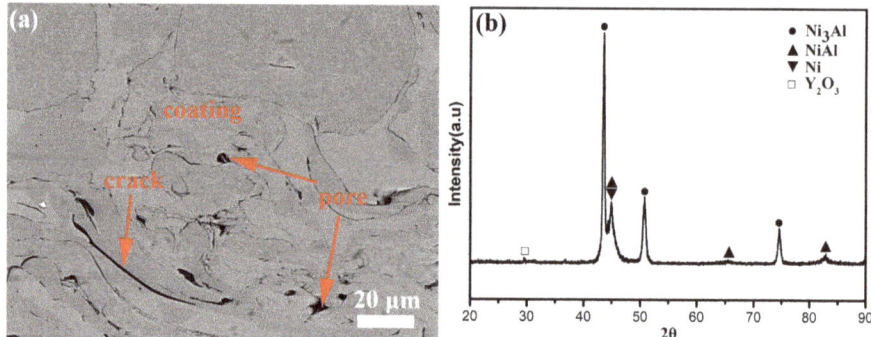

Figure 3. SEM morphology of (**a**) cross-section and (**b**) XRD pattern of NiCrAlY coating.

3.2. Electrochemical Performance of NiCrAlY Coating

Figure 4 gives the potentiodynamic polarization curves of NiCrAlY coating sliding conditions in different corrosive solutions. Key test parameters such as the corrosion potential (E_{corr}), corrosion current density (i_{corr}), anodic and cathodic Tafel slopes (β_a and β_c) are obtained from Figure 4 and shown in Table 1. The polarization resistance value (R_p) is calculated by Stern–Geary equation:

$$R_p = \frac{\beta_a \times \beta_c}{2.303 i_{corr}(\beta_a + \beta_c)}. \tag{1}$$

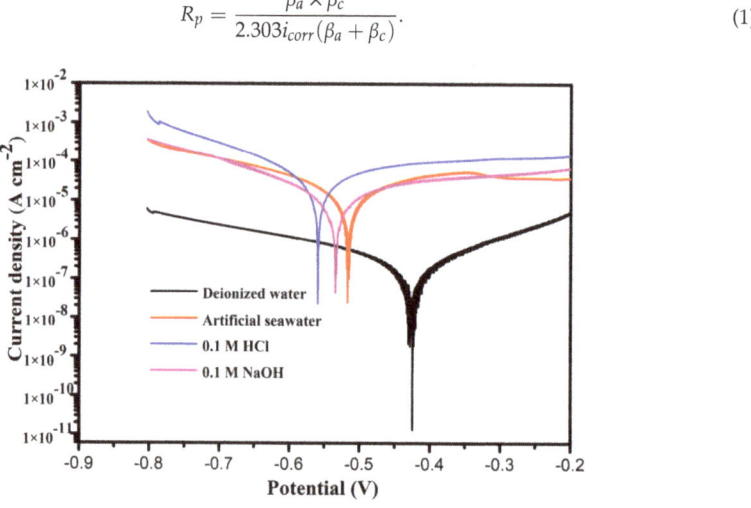

Figure 4. Potentiodynamic polarization curves of NiCrAlY coating sliding conditions in different corrosive solutions.

Results indicate that the corrosion potential (E_{corr}) of NiCrAlY coating under deionized water is the highest of −0.428 V (vs. SCE). However, the E_{corr} of the coating in artificial seawater, HCl and NaOH shift to −0.516 V (vs. SCE), −0.559 V (vs. SCE) and −0.535 V (vs. SCE) respectively. Simultaneously, the corrosion current density (i_{corr}) of this coating in deionized water shows the lowest of 7.865×10^{-8} A/cm^2. Generally speaking, corrosion current density, whose rate is often used as corrosion rate, is a crucial reference to evaluate corrosion resistance [13,27]. Therefore, the coating under deionized water with the lowest corrosion rate is hard to corrode. The coating in HCl presenting the highest corrosion current density is very easy to be corroded. At the same time, the coating in deionized water has the highest β_a, β_c and R_p of 0.072 V/dec, 0.049 V/dec and 1.610×10^5 Ω respectively, which further illustrates that the coating in deionized water holds a good corrosion resistance.

Table 1. Corrosion parameters of NiCrAlY coating from potentiodynamic polarization curves.

Corrosive Solutions	E_{corr} (V, vs. SCE)	i_{corr} (A/cm²)	β_a (V/dec)	$-\beta_c$ (V/dec)	R_p (Ω)
Deionized water	−0.428	7.865×10^{-8}	0.072	0.049	1.610×10^5
Artificial seawater	−0.516	8.986×10^{-7}	0.043	0.042	1.027×10^4
0.1 M HCl	−0.559	3.356×10^{-6}	0.036	0.039	2.422×10^3
0.1 M NaOH	−0.535	1.039×10^{-6}	0.039	0.038	8.044×10^3

3.3. Tribological Behavior of NiCrAlY Coating

Figure 5 shows the friction curves and wear rate of NiCrAlY coating in different corrosive solutions. The friction coefficient (COF) of the coating under the NaOH solution was the highest, with a value of 0.46. In artificial seawater and deionized water, it was 0.37 and 0.26, respectively. Surprisingly, the COF reduced to 0.24 and remains steady in HCl solution. Nevertheless, the NiCrAlY coating has a high wear rate (WR) of 6.36×10^{-6} mm³/Nm in the HCl solution. This phenomenon is likely to show the high corrosion rate of coating in HCl solution (Figure 4). The synergistic effect of corrosion and wear in a corrosive environment leads to the loss of large material, which usually larger than the synergistic effect of the sum of corrosion and wear [28,29]. So, the coating under the HCl solution presents a more obvious wear rate. The coating in the NaOH solution has the highest wear rate of 6.89×10^{-6} mm³/Nm. At the same time, the coating in deionized water gives the lowest WR of 2.36×10^{-6} mm³/Nm, which is caused by the lowest corrosion rate of coating in deionized water (Figure 4).

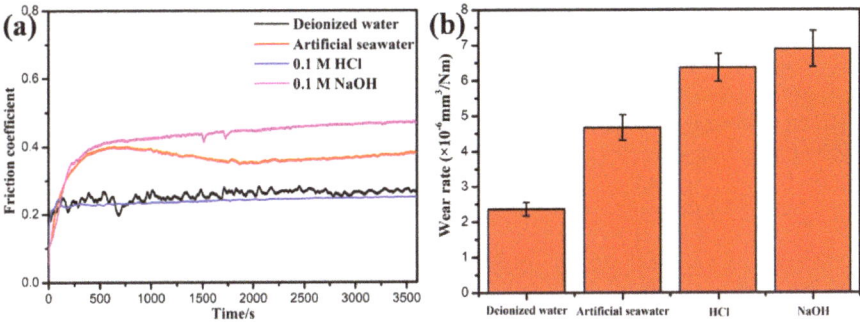

Figure 5. Friction curves (a) and wear rate (b) of NiCrAlY coating in different corrosive solutions.

Figure 6 presents the 2D and 3D configurations of NiCrAlY coating worn surfaces in different corrosive solutions. The worn surface has the shallowest and narrowest friction trace in deionized water (Figure 6a,e). Therefore, the COF and WR are lower (Figure 5). It further illustrates that the coating in deionized water shows excellent corrosion and wear resistance. The worn surface of NiCrAlY coating in HCl corrosive solution is very rough and has serious corrosion (Figure 6c). So, the coating obtains high WR under HCl corrosive solution (Figure 5). The worn track of NiCrAlY coating in NaOH corrosive solution is the deepest and widest (Figure 6d,f). Therefore, this coating has the worst tribological performance (Figure 5).

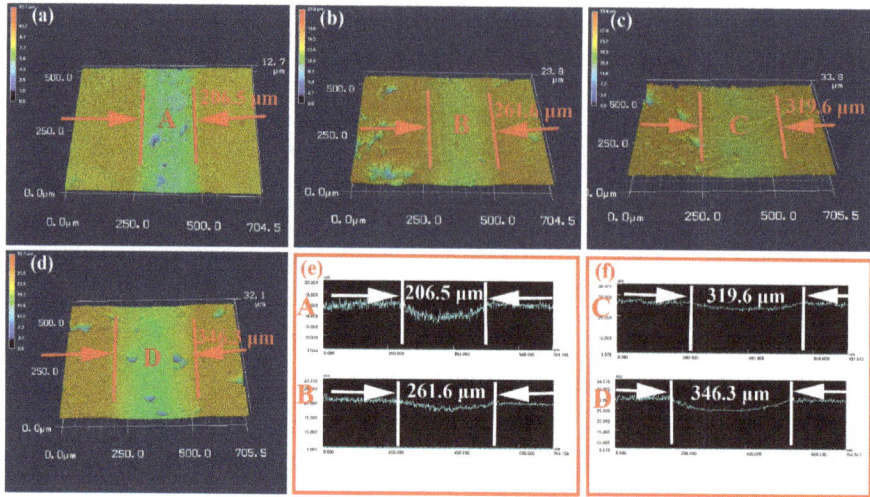

Figure 6. 2D and 3D configurations of NiCrAlY coating worn surfaces in different corrosive solutions: (**a**) deionized water, (**b**) artificial seawater, (**c**) HCl solution and (**d**) NaOH solution; (**e**) 2D profiles of A and B regions; (**f**) 2D profiles of C and D regions.

To further research the influence of corrosive solution upon the corrosion-wear property of NiCrAlY coating, Raman analysis is tested. Figure 7 shows the Raman spectra of the worn surface of NiCrAlY coating in different corrosive solutions. The Al_2O_3, Cr_2O_3 and NiO are the main phases on the worn surface of NiCrAlY coating after sliding in deionized water, artificial seawater and NaOH solution. Nevertheless, the worn surface of NiCrAlY coating observes the new phase of $Al(OH)_3$ after sliding in HCl corrosive solution [30]. The results indicate that the NiCrAlY coating has suffered serious corrosion in the HCl corrosive solution because of the existence of stronger and more numerous peaks [1]. The corrosion products are easily worn out during the friction process. So the wear rate of the coating under HCl solution is very high (Figure 5).

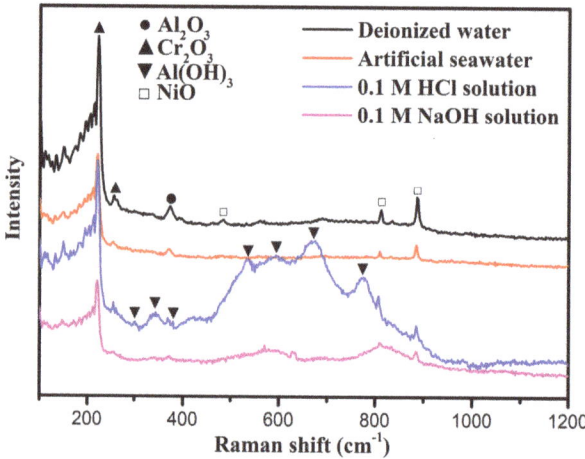

Figure 7. Raman spectra of worn surface of NiCrAlY coating in different corrosive solutions.

3.4. Lubrication Behavior of Al(OH)$_3$ on NiCrAlY Coating in HCl Solution

Figure 8 shows the corrosion-wear mechanisms of NiCrAlY coating in the HCl solution. The surface becomes very smooth because the corrosion-wear effect with the mix of oxides and hydroxides formed by electrochemical reactions (Figures 6 and 7). In terms of the potential values, aluminum is the least noble element and the order of potentials follows Ni > Cr > Al [1]. So the aluminum element is more likely to be corroded at first. The following electrochemical reactions could explain the process of Al(OH)$_3$ formation:

$$Al \rightarrow Al^{3+} + 3e^- \tag{2}$$

$$2H^+ + 2e^- \rightarrow H_2 \tag{3}$$

$$H_2O + 2e^- \rightarrow H_2 + 2OH^- \tag{4}$$

$$Al^{3+} + 3OH^- \rightarrow Al(OH)_3 \tag{5}$$

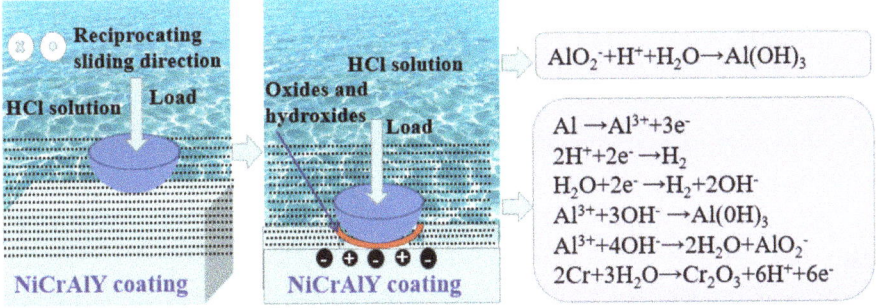

Figure 8. Schematic diagram of corrosion-wear mechanisms of NiCrAlY coating in HCl solution.

Terryn et al. [31] illustrated that the generation of Al(OH)$_3$ is related to local pH changes in the hydrogen reduction region. Hence, the Al(OH)$_3$ could be formed where the hydrogen evolution occurs. Furthermore, when the local pH rises to above 9, Al^{3+} ions will react with excessive OH$^-$ ions and forms aluminate anions [1]. Aluminate anions cannot maintain stable in HCl corrosive solution and will precipitate as Al(OH)$_3$ (Figure 8). This reaction can be described as follows:

$$Al^{3+} + 4OH^- \rightarrow AlO_2^- + 2H_2O \tag{6}$$

$$AlO_2^- + H^+ + H_2O \rightarrow Al(OH)_3 \tag{7}$$

Thus, it inexistences the Al(OH)$_3$ on the worn surface of NiCrAlY coating in NaOH solution in the process of sliding but the following reaction [32]:

$$2Al + 2OH^- + H_2O \rightarrow 2AlO_2^- + 2H_2 \tag{8}$$

Of course, in addition to the Al dissolution, according to the standard of electrode potentials, Cr element is also dissolved at the anodic cycle and is electrochemically oxidized to Cr$_2$O$_3$, which is well consistent with the micro-Raman results (Figure 7) [1]. The oxidation reaction process can be illustrated as follows [30]:

$$2Cr + 3H_2O \rightarrow Cr_2O_3 + 6H^+ + 6e^- \tag{9}$$

The above oxidation reactions and metal dissolution explain the smooth surface. Al(OH)$_3$ can be evenly distributed on the smooth worn surface and effectively reduce the direct contact of counter ball and sample. At the same time, the frictional shear stress can form the lubricating layer on the worn surface, which can obviously reduce the friction coefficient of coating in HCl corrosive solution [1]. Therefore, the COF of NiCrAlY coating in the HCl corrosive solution is the lowest of 0.24 (Figure 5).

4. Conclusions

In this work, the corrosion-wear properties of NiCrAlY coating were studied under deionized water, artificial seawater, 0.1 M HCl solution and 0.1 M NaOH. The main conclusions are given as follows:

(1) The NiCrAlY coating is composed of Ni$_3$Al, nickel-based solid solution, NiAl and Y$_2$O$_3$.
(2) In deionized water, the NiCrAlY coating with the lowest corrosion current density of 7.865×10^{-8} A/cm^2 is hard to erode. Meanwhile, it presents a lower friction coefficient and the lowest wear rate.
(3) In HCl corrosive solution, the coating gives the highest corrosion current density (i_{corr}) of 3.356×10^{-6} A/cm^2 and a higher wear rate of 6.36×10^{-6} mm^3/Nm.
(4) In HCl corrosive solution, the emergence of Al(OH)$_3$ on the coating surface could reduce the direct contact between the counter ball and sample effectively, which is conducive to the lowest friction coefficient of 0.24.

Author Contributions: Data curation, Y.K.; Formal analysis, S.Z.; Project administration, Y.G.; Resources, Q.Z.; Software, Y.L.; Visualization, H.G.; Writing—original draft, B.L.; Writing—review & editing, C.L. All authors have read and agreed to the published version of the manuscript.

Funding: This research was funded by the National Natural Science Foundation of China (Grant No. 51805408, 51665026), the Natural Science Foundation of Shaanxi Province (Grant No. 2019JQ-283), the China Postdoctoral Science Foundation (Grant No. 2019M653597), the Shaanxi Province Postdoctoral Science Foundation, the Fundamental Research Funds for Central Universities (Grant No. xzy012019010, xtr0118008), the Guangxi Innovation Driven Development Project (Grant No. GUIKEAA18242001) and the Guangdong Province Key Area R&D Program (Grant No. 2019B010942001).

Conflicts of Interest: The authors declare no conflict of interest.

References

1. Liu, X.; An, Y.; Li, S.; Zhao, X.; Hou, G.; Zhou, H.; Chen, J. An assessment of tribological performance on NiCoCrAlYTa coating under corrosive environments. *Tribol. Int.* **2017**, *115*, 35–44. [CrossRef]
2. Ma, F.; Li, J.; Zeng, Z.; Gao, Y. Structural, mechanical and tribocorrosion behaviour in artificial seawater of CrN/AlN nano-multilayer coatings on F690 steel substrates. *Appl. Surf. Sci.* **2018**, *428*, 404–414. [CrossRef]
3. Huttunen-Saarivirta, E.; Kilpi, L.; Hakala, T.J.; Carpen, L.; Ronkainen, H. Tribocorrosion study of martensitic and austenitic stainless steels in 0.01 M NaCl solution. *Tribol. Int.* **2016**, *95*, 358–371. [CrossRef]
4. Wang, C.; Ye, Y.; Guan, X.; Hu, J.; Wang, Y.; Li, J. An analysis of tribological performance on Cr/GLC film coupling with Si$_3$N$_4$, SiC, WC, Al$_2$O$_3$ and ZrO$_2$ in seawater. *Tribol. Int.* **2016**, *96*, 77–86. [CrossRef]
5. Liu, X.; An, Y.; Zhao, X.; Li, S.; Deng, W.; Hou, G.; Ye, Y.; Zhou, H.; Chen, J. Hot corrosion behavior of NiCoCrAlYTa coating deposited on Inconel alloy substrate by high velocity oxy-fuel spraying upon exposure to molten V$_2$O$_5$-containing salts. *Corros. Sci.* **2016**, *112*, 696–709. [CrossRef]
6. Meng, Y.; Su, F.; Chen, Y. Nickel/Multi-walled Carbon Nanotube Nanocomposite Synthesized in Supercritical Fluid as Efficient Lubricant Additive for Mineral Oil. *Tribol. Lett.* **2018**, *66*, 134. [CrossRef]
7. Pereira, J.; Zambrano, J.; Afonso, C.; Amigo, V. Microstructure and mechanical properties of NiCoCrAlYTa alloy processed by press and sintering route. *Mater. Charact.* **2015**, *101*, 159–165. [CrossRef]
8. Anghel, E.; Marcu, M.; Banu, A.; Atkinson, I.; Paraschiv, A.; Petrescu, S. Microstructure and oxidation resistance of a NiCrAlY/Al$_2$O$_3$-sprayed coating on Ti-19Al-10Nb-V alloy. *Ceram. Int.* **2016**, *42*, 12148–12155. [CrossRef]

9. Bolelli, G.; Candeli, A.; Lusvarghi, L.; Ravaux, A.; Cazes, K.; Denoirjean, A.; Valette, S. Tribology of NiCrAlY+Al$_2$O$_3$ composite coatings by plasma spraying with hybrid feeding of dry powder + suspension. *Wear* **2015**, *344–345*, 69–85. [CrossRef]
10. Eriksson, R.; Yuan, K.; Li, X.; Peng, R. Corrosion of NiCoCrAlY Coatings and TBC Systems Subjected to Water Vapor and Sodium Sulfate. *J. Therm. Spray Technol.* **2015**, *24*, 953–964. [CrossRef]
11. Habib, K.; Damra, M.; Carpio, J.; Cervera, I.; Saura, J. Performance of NiCrAlY Coatings Deposited by Oxyfuel Thermal Spraying in High Temperature Chlorine Environment. *J. Mater. Eng. Perform.* **2014**, *23*, 3511–3522. [CrossRef]
12. Li, B.; Gao, Y.; Jia, J.; Han, M.; Guo, H.; Wang, W. Influence of heat treatments on the microstructure as well as mechanical and tribological properties of NiCrAlY-Mo-Ag coatings. *J. Alloy. Compd.* **2016**, *686*, 503–510. [CrossRef]
13. Liu, X.; Zhao, X.; An, Y.; Hou, G.; Li, S.; Deng, W.; Zhou, H.; Chen, J. Effects of loads on corrosion-wear synergism of NiCoCrAlYTa coating in artificial seawater. *Tribol. Int.* **2018**, *118*, 421–431. [CrossRef]
14. Peng, X.; Jiang, S.; Gong, J.; Sun, X.; Sun, C. Preparation and Hot Corrosion Behavior of a NiCrAlY + AlNiY Composite Coating. *J. Mater. Sci. Technol.* **2016**, *32*, 587–592. [CrossRef]
15. Yaghtin, A.; Javadpour, S.; Shariat, M. Hot corrosion of nanostructured CoNiCrAlYSi coatings deposited by high velocity oxy fuel process. *J. Alloy. Compd.* **2014**, *584*, 303–307. [CrossRef]
16. Bakhsheshi-Rad, H.; Hamzah, E.; Ismail, A.; Daroonparvar, M.; Yajid, M.; Medraj, M. Preparation and characterization of NiCrAlY/nano-YSZ/PCL composite coatings obtained by combination of atmospheric plasma spraying and dip coating on Mg-Ca alloy. *J. Alloy. Compd.* **2016**, *65*, 440–452. [CrossRef]
17. Demian, C.; Denoirjean, A.; Pawlowski, L.; Denoirjean, P.; Ouardi, R. Microstructural investigations of NiCrAlY+Y$_2$O$_3$ stabilized ZrO$_2$ cermet coatings deposited by plasma transferred arc (PTA). *Surf. Coat. Technol.* **2016**, *300*, 104–109. [CrossRef]
18. Li, B.; Jia, J.; Han, M.; Gao, Y.; Wang, W.; Li, C. Microstructure, mechanical and tribological properties of plasma-sprayed NiCrAlY-Mo-Ag coatings from conventional and nanostructured powders. *Surf. Coat. Technol.* **2017**, *324*, 552–559. [CrossRef]
19. Liu, Y.; Hu, X.; Zheng, S.; Zhu, Y.; Wei, H.; Ma, X. Microstructural evolution of the interface between NiCrAlY coating and superalloy during isothermal oxidation. *Mater. Des.* **2015**, *80*, 63–69. [CrossRef]
20. Tahari, M.; Shamanian, M.; Salehi, M. Microstructural and morphological evaluation of MCrAlY/YSZ composite produced by mechanical alloying method. *J. Alloy. Compd.* **2012**, *525*, 44–52. [CrossRef]
21. Wang, J.; Chen, M.; Yang, L.; Zhu, S.; Wang, F. Comparative study of oxidation and interdiffusion behavior of AIP NiCrAlY and sputtered nanocrystalline coatings on a nickel-based single-crystal superalloy. *Corros. Sci.* **2015**, *98*, 530–540. [CrossRef]
22. Chen, J.; Cheng, J.; Zhu, S.; Tan, H.; Qiao, Z.; Yang, J. Tribological Behaviors of Cu/AlMgB$_{14}$ Composite Under Deionized Water and Liquid Paraffin. *Tribol. Lett.* **2018**, *67*. [CrossRef]
23. Espallargas, N.; Mischler, S. Tribocorrosion behaviour of overlay welded Ni-Cr 625 alloy in sulphuric and nitric acids: Electrochemical and chemical effects. *Tribol. Int.* **2010**, *43*, 1209–1217. [CrossRef]
24. Shan, L.; Wang, Y.; Zhang, Y.; Zhang, Q.; Xue, Q. Tribocorrosion behaviors of PVD CrN coated stainless steel in seawater. *Wear* **2016**, *362–363*, 97–104. [CrossRef]
25. Chen, J.; An, Y.; Yang, J.; Zhao, X.; Yan, F.; Zhou, H.; Chen, J. Tribological properties of adaptive NiCrAlY-Ag-Mo coatings prepared by atmospheric plasma spraying. *Surf. Coat. Technol.* **2013**, *235*, 521–528. [CrossRef]
26. Cai, B.; Tan, Y.; Tan, H.; Jing, Q.; Zhang, Z. Tribological behavior and mechanism of NiCrBSi-Y$_2$O$_3$ composite coatings. *Trans. Nonferrous Met. Soc. China* **2013**, *23*, 2002–2010. [CrossRef]
27. Zhang, R.; Wang, H.; Xing, X.; Yuan, Z.; Yang, S.; Han, Z.; Yuan, G. Effects of Ni addition on tribocorrosion property of TiCu alloy. *Tribol. Int.* **2017**, *107*, 39–47. [CrossRef]
28. Buciumeanu, M.; Bagheri, A.; Souza, J.; Silva, F.; Henriques, B. Tribocorrosion behavior of hot pressed CoCrMo alloys in artificial saliva. *Tribol. Int.* **2016**, *97*, 423–430. [CrossRef]
29. Mischler, S. Triboelectrochemical techniques and interpretation methods in tribocorrosion: A comparative evaluation. *Tribol. Int.* **2008**, *41*, 573–583. [CrossRef]
30. Huang, E.; Li, A.; Xu, J.; Chen, R.; Yamanaka, T. High-pressure phase transition in Al(OH)$_3$: Raman and X-ray observations. *Geophys. Res. Lett.* **1996**, *23*, 3083–3086. [CrossRef]

31. Terryn, H.; Vereecken, J.; Thompson, G. The electrograining of aluminium in hydrochloric acid-II. Formation of ETCH products. *Corros. Sci.* **1991**, *32*, 73–88. [CrossRef]
32. Wang, Y.; Kong, G.; Che, C. Corrosion behavior of Zn-Al alloys in saturated Ca(OH)$_2$ solution. *Corros. Sci.* **2016**, *112*, 679–686. [CrossRef]

© 2020 by the authors. Licensee MDPI, Basel, Switzerland. This article is an open access article distributed under the terms and conditions of the Creative Commons Attribution (CC BY) license (http://creativecommons.org/licenses/by/4.0/).

Article

Improvement of Mechanical Properties of Plasma Sprayed Al₂O₃–ZrO₂–SiO₂ Amorphous Coatings by Surface Crystallization

Jan Medricky [1,2,*], Frantisek Lukac [1], Stefan Csaki [1], Sarka Houdkova [3], Maria Barbosa [4], Tomas Tesar [1,2], Jan Cizek [1], Radek Musalek [1], Ondrej Kovarik [2] and Tomas Chraska [1]

[1] Institute of Plasma Physics, The Czech Academy of Sciences, Za Slovankou 1782/3, Prague 182 00, The Czech Republic; lukac@ipp.cas.cz (F.L.); csaki@ipp.cas.cz (S.C.); tesar@ipp.cas.cz (T.T.); cizek@ipp.cas.cz (J.C.); musalek@ipp.cas.cz (R.M.); chraskat@ipp.cas.cz (T.C.)
[2] Department of Materials, Faculty of Nuclear Sciences and Physical Engineering, Czech Technical University in Prague, Prague 115 19, The Czech Republic; ondrej.kovarik@fjfi.cvut.cz
[3] Research and Testing Institute Pilzen, Pilsen 301 00, The Czech Republic; houdkova@vzuplzen.cz
[4] Fraunhofer IWS, 01277 Dresden, Germany; maria.barbosa@iws.fraunhofer.de
* Correspondence: medricky@ipp.cas.cz; Tel.: +420-266-052-931

Received: 14 August 2019; Accepted: 26 September 2019; Published: 2 October 2019

Abstract: Ceramic Al_2O_3–ZrO_2–SiO_2 coatings with near eutectic composition were plasma sprayed using hybrid water stabilized plasma torch (WSP-H). The as-sprayed coatings possessed fully amorphous microstructure which can be transformed to nanocrystalline by further heat treatment. The amorphous/crystalline content ratio and the crystallite sizes can be controlled by a specific choice of heat treatment conditions, subsequently leading to significant changes in the microstructure and mechanical properties of the coatings, such as hardness or wear resistance. In this study, two advanced methods of surface heat treatment were realized by plasma jet or by high energy laser heating. As opposed to the traditional furnace treatments, inducing homogeneous changes throughout the material, both approaches lead to a formation of gradient microstructure within the coatings; from dominantly amorphous at the substrate–coating interface vicinity to fully nanocrystalline near its surface. The processes can also be applied for large-scale applications and do not induce detrimental changes to the underlying substrate materials. The respective mechanical response was evaluated by measuring coating hardness profile and wear resistance. For some of the heat treatment conditions, an increase in the coating microhardness by factor up to 1.8 was observed, as well as improvement of wear resistance behaviour up to 6.5 times. The phase composition changes were analysed by X-ray diffraction and the microstructure was investigated by scanning electron microscopy.

Keywords: amorphous; nanocrystalline; wear resistant; Vickers microhardness; plasma spraying

1. Introduction

Thermally sprayed ceramic coatings are widely used in industry to provide mechanical, chemical and thermal protection. Coatings are prepared by introducing the feedstock material, most often in a form of powder, into a hot plasma jet, where it is melted and propelled towards a prepared substrate. After their impact at the substrate, the molten particles flatten and solidify in a form of disk-like platelets called splats. Plasma spraying inherently possesses extremely high cooling rates of the particles, in the range 10^3–10^6 K/s, thereby frequently giving rise to a formation of non-equilibrium phases, microstructures of fine columnar grains [1] or even amorphous phases [2].

Due to its low price as well as good chemical and wear resistance, Al_2O_3 is often used as a material of the first choice for protecting metallic parts from wear and corrosion. Additionally, Al_2O_3

properties can be further significantly improved when it is mixed with other components. For example, $Al_2O_3-Y_2O_3$ and $Al_2O_3-TiO_2$ composite coatings possess higher wear resistance than pure Al_2O_3 [3,4], while the addition of ZrO_2 increases the coatings' toughness [5].

Another way to improve the coating properties is a preparation of nanocrystalline or sub-micron microstructure: nanocrystalline materials are characterized by a microstructural length or grain size of up to about 100 nm, while microstructure having grain sizes from ~0.1 to 0.3 µm are classified as submicron materials [6]. It was shown that nanocrystalline materials possess better mechanical properties than their coarse-grained counterparts [7]. For instance, a nanocrystalline Inconel 718 coating deposited by high velocity oxygen fuel (HVOF) method exhibited a significant increase in hardness (by approximately 60%) over that of the Inconel 718 control sample [8]. Another results published in [9] showed that decreasing the grain size of Al_2O_3 feedstock powder from ~50 µm down to 300 nm increased the tensile adhesion strength of the deposited coating by a factor of three and the coating wear resistance was increased by a factor of ten. Nanostructured zirconia coatings deposited by plasma spraying in the study by Chen et al. [10] showed that the wear rates of the nanostructured coatings were about 40% of those of traditional zirconia coatings under loads from 20 to 80 N. Owing to these outstanding performances, targeted applications have been successfully implemented for hard and wear-resistant ceramic coatings in industrial sectors in the past decade [6].

Suspension plasma spraying has been intensively studied as a reliable method of preparation of coatings with such fine, nanometric-grain microstructure. In the early suspension experiments, nano-sized particles were used. However, their tendency towards agglomeration as well as associated difficult handling and potential health risk of using such suspensions [11] triggered a shift to using sub-micrometric particles instead. Even though the coatings prepared by suspension plasma spraying route can often surpass the coatings prepared from dry powders [9], their industrial application remain rather scarce at the moment, owing to difficulties in the coating preparation and relatively low deposition rate (coating thickness increase per torch pass) compared to dry powder plasma spraying [12,13].

An alternative approach to preparation of nanocrystalline coatings is based on deposition of amorphous coatings from coarse powders and their subsequent heat treatment in order to induce growth of nanocrystalline grains in the microstructure. Deposition of such amorphous microstructures can be relatively easily implemented through the rapid solidification of the particles in plasma spraying, provided the cooling rates are sufficiently high to fully suppress crystalline growth. It has been previously reported that materials with near-eutectic composition can solidify as fully or partially amorphous solids, and this finding was widely used for preparation of e.g. metallic glasses [14] or amorphous ceramics, such as $Al_2O_3-Y_2O_3$, $Al_2O_3-ZrO_2$ and $Al_2O_3-ZrO_2-SiO_2$ systems, as presented in [3,15–19]. Transformation of amorphous coating of the ternary $Al_2O_3-ZrO_2-SiO_2$ system into nanocrystalline coating was successfully reported in our previous studies [19,20], where an improvement of the mechanical properties was also described. In both studies, amorphous atmospheric plasma sprayed materials $Al_2O_3-ZrO_2$ and $Al_2O_3-ZrO_2-SiO_2$ were isothermally heat treated at various temperatures above the crystallization temperature (980 °C), forming nanocrystallites embedded in the amorphous matrix. Size of the nanocrystallites was strongly dependent on the heat treatment temperature and resulted in different mechanical properties, such as hardness and flexural strength [20].

The traditional way of heat treatment of amorphous materials would be furnace treatment; however, it is not applicable for coatings on metallic substrates because of substrate oxidation, grain growth and even possible spallation of the coating, caused by the differences in the substrate–coating coefficients of thermal expansion. In this paper, we present two alternative methods of heat treatment of amorphous $Al_2O_3-ZrO_2-SiO_2$ coatings deposited by atmospheric hybrid water stabilized plasma torch (WSP-H). The study is focused on laser and plasma surface heat treatment techniques, i.e., industrially relevant methods with high potential. As opposed to furnace annealing (used as a reference method for free standing ceramic parts), both of these methods enable on-site modification of the coating properties without inducing any detrimental changes to the metallic substrate material and enable surface heat treatment of large-sized components, such as paper-mill rollers. Upon the treatment, the phase composition and microstructure changes were evaluated by XRD and SEM, respectively, and the associated mechanical response of the coating to heat treatment was studied by measuring the Vickers microhardness and Pin on Disc wear resistance tests.

2. Materials and Methods

Feedstock Al_2O_3–ZrO_2–SiO_2 ceramic powder was obtained by crushing commercially available bulk material Eucor (Eutit Ltd., Stara Voda, Czech Republic). The powder was sieved into a sprayable size distribution with D_{50} = 89 µm, as measured by particle size analyser Mastersizer 3000 (Malvern, UK). Using the EDX analysis, the near eutectic composition of the feedstock powder was determined as 49% Al_2O_3, 31% ZrO_2 and 19% SiO_2 (all ± 2%). The hybrid water stabilized plasma torch WSP-H 500 (ProjectSoft, Hradec Kralove, Czech Republic) was used for plasma spraying onto S235 steel substrates with dimensions 50 × 30 × 10 mm^3. The torch was operated at 500 A (~150 kW) and 15 slpm argon flow rate. In addition to argon, the torch consumes about 20 g/min of demineralized water, which is evaporated and ionized to supply the plasma with hydrogen and oxygen ions (for detailed information about spraying with the WSP-H torch, refer to [21,22]). The stand-off distance was set to 350 mm and powder was injected radially into plasma jet at a feeding distance (i.e., the distance of the powder injection point from the torch exit nozzle) of 35 mm and powder feed rate of 10 kg/h (167 g/min).

Prior to the deposition, the substrates were grit blasted by alumina grit (Ra = 8.1 ± 0.3 µm) and mounted to a revolving carousel. The substrates temperature during the deposition was measured by infra-red camera TIM160 (Micro-Epsilon, Ortenburg, Germany) facing the substrates' front side, as well as by a K-type thermocouple inserted into a hole drilled from the back-side of one of the samples and reaching 1 mm under the coated surface. The substrates were preheated by 3 cycles of plasma torch with deactivated powder feeding. To prepare a coating with a 1.5 mm thickness, 11 successive plasma torch cycles were needed. One deposition cycle consisted of three up and down strokes and was followed by an extensive cooling. Each deposition cycle was manually triggered when temperature measured by the thermocouple dropped to 250 °C.

To facilitate an accurate determination of the heat treatment variants' influence, the as-sprayed coatings were polished down using a P600 diamond disc. Subsequently, the samples were subjected to one of two methods of surface heat-treatment: laser or plasma. The laser treatment was performed using a high power diode laser (Laserline GmbH, Muhlheim-Karlich, Germany) with maximal output power of 9 kW, λ = 915–1030 nm, 1000 µm fibre diameter, 400 mm focal length and laser focus diameter 7.5 mm. The laser transverse velocity and power were varied to obtain 15 different laser heat-treatment conditions in total. On the other hand, high enthalpy plasma generated heat produced by the plasma torch offers a quick and readily available alternative since the heat treatment can be performed directly after the coating deposition using the same plasma torch that was used for the coating deposition. The torch transverse velocity and power were modified, in order to prepare 6 different plasma heat-treatment conditions. The samples were mounted into a stationary sample holder and heat-treated by a single pass of the plasma torch at a stand-off distance of 150 mm. A schematic illustration of the two surface heat-treatment methods is provided in Figure 1.

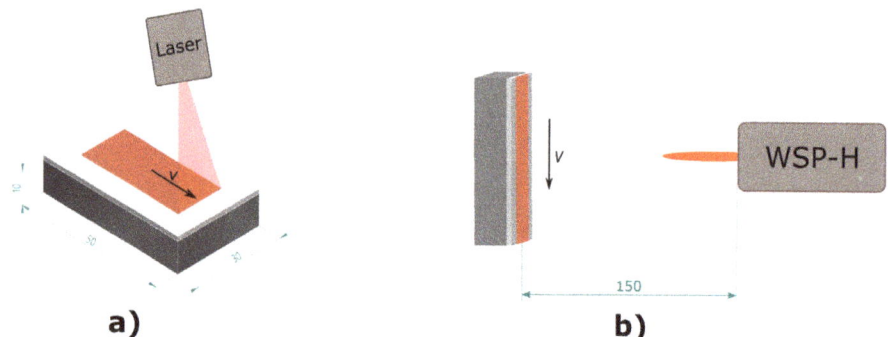

Figure 1. Schematic illustration of samples surface heat-treatment by laser (**a**) and WSP-H plasma torch (**b**).

Stripped-off ceramic coatings were then also prepared from the as-sprayed samples by grinding off the substrates. These coatings were used for measurement of the thermal expansion using vertical dilatometer Setsys 16/18 (Setaram, Caluire-et-Cuire, France) and to determine the crystallization onset temperatures using a Bahr STA 504 differential thermal analyser (Bahr, Hullhorst, Germany). In addition to the two surface heat-treatment methods, complimentary furnace annealing of the stripped-off ceramic samples was carried out. The furnace Entech EEF 5/16-HV (Entech, Angelholm, Sweden) was first preheated to the temperature 1050 °C and the samples were then inserted for a specified time. Such samples, isothermally treated in the whole volume (as opposed to gradient heating during laser or plasma treatment), were used as a reference set.

Metallographic samples of all specimens were prepared using Tegramin-25 automated polishing system (Struers, Willich, Denmark). The polished cross-sections were observed using a scanning electron microscope EVO MA 15 (Carl Zeiss SMT, Oberkochen, Germany) equipped with XFlash 5010 energy-dispersive spectrometer EDX (Bruker, Hamburg, Germany). Porosity of the coating was evaluated from seven SEM micrographs with nominal magnification 500× using semi-automatic thresholding procedure in ImageJ software (National Institutes of Health, Bethesda, MD, USA). Vickers microhardness profiles were measured on the polished sample cross-sections throughout the coating thickness using Q10A+ universal hardness tester (Qness, Golling an der Salzach, Austria), using the load of 300 g and dwell time 10 s. The average value of Vickers microhardness was calculated from at least 5 indents. Phase composition was evaluated on the free surfaces of the samples by powder X-ray diffractometer (XRD) D8 Discover (Bruker, Hamburg, Germany), using Cu anode and equipped with 1D detector. The degree of crystallinity, size of coherently diffracting domains (CDD) and microstrains were evaluated by quantitative Rietveld analysis of the acquired XRD spectra. Broadening of the diffraction peaks and background fitting were analysed using TOPAS V5 software (Bruker AXS, Hamburg, Germany). It was assumed that the effects of small crystallite size and microstrains contribute to broadening of Lorentzian and Gaussian components of pseudo-Voigt function, respectively [23].

Tribological properties were measured by CSEM High Temperature Tribometer (Anton Paar GmbH, Graz, Austria) by dry sliding Pin on Disc test according to ASTM G99 05 standard. The tests were carried out at room temperature, in air atmosphere (31% relative humidity) without lubrication using alumina counterpart ball (6 mm diameter) with 10 N normal load, 0.1 m·s^{-1} speed and measured distance of 110 m in 5000 cycles (track radius 3.5 mm). The wear tracks profiles were measured by profilometer P-6 Profiler (KLA-Tencor, Milpitas, CA, USA), at four different places, and the wear volume was calculated.

3. Results

3.1. As-Sprayed Samples

Cross-sections, prepared from the as-sprayed coating, were analysed using SEM in Back Scattered Electron (BSE) mode to observe the microstructure and overall coating quality. As seen from Figure 2a, the coatings evenly covered the substrate and adhered well to it with neither delaminations nor vertical cracks observed. The average chemical composition of the coating in wt.%, as evaluated by EDX, was: 51 ± 1 Al_2O_3, 33 ± 2 ZrO_2, 13 ± 1 SiO_2, with traces of Fe- and Na-oxides. From the magnified view in Figure 2b, it can be observed that individual splats differed significantly in the shade of grey, which is caused by variations of their chemical composition. The brighter the splat, the more ZrO_2 it contains, while darker splats are richer in Al_2O_3, as was confirmed by EDX analysis (see Figure 3). Apart from the compositional variations of the splats, some unmelted feedstock particles were observed embedded in the as-sprayed coating. These particles can be easily recognized by their original eutectic microstructure (Figure 2b), which was retained from the feedstock powder. The amount of unmelted particles within the coating was 4.4 ± 0.9%, as evaluated by the image analysis. The magnified view also shows short micro cracks within the coating, with an average length of about 80 µm. The total porosity of the as-sprayed coating, as evaluated by the image analysis, was 4.7 ± 0.2% and consisted mainly of globular pores with average size around 8 µm².

The XRD measurement of the samples free surfaces showed that the as-sprayed coatings were mainly amorphous, with only about 8% of crystalline phases present, assumedly dominantly formed by unmelted particles as described above.

3.2. Thermal Properties

Free standing ceramic samples were prepared from the as-sprayed coatings by grinding off the substrate. These samples were used for Differential Thermal Analysis (DTA) to obtain crystallization temperature of the amorphous samples. The DTA showed onset crystallization peak at 984 °C, and the crystallization was fully finished at 1015 °C, as shown in Figure 4. Moreover, thermal dilatometry was measured suggesting a rapid linear shrinkage of 2.47% observed at crystallization temperature (blue dash-and-dot line in Figure 4). The average value of the coefficient of thermal expansion (CTE) before crystallization, determined from measurement of displacement, was (3.5 ± 0.7) × 10^{-6} K^{-1} and changed to (6.1 ± 0.4) × 10^{-6} K^{-1} after crystallization. When the identical sample was subjected to a second measurement of displacement, CTE remained constant within the full temperature range up to 1300 °C, suggesting that the primary crystallization is an irreversible transformation.

3.3. Heat Treatment

The as-sprayed samples were surface heat-treated by laser or plasma. Additionally, two free standing ceramic samples were furnace heat-treated at 1050 °C (i.e., slightly above the determined crystallization temperature) with dwell time 1 and 5 min, for further comparison with the surface heat-treated samples. Parameters of all heat treatment conditions are listed in Table 1. Notation of the samples in Table 1 is as follows: AS—as sprayed sample, F#—Furnace, L#—Laser and P#—Plasma heat treated samples. Please consider that plasma torch used in the experiment had power of 100–150 kW. Therefore, to prevent melting of the samples, the transverse velocities used for plasma torch treatment had to be significantly higher than the ones used for laser treatment.

The heat treatment of the samples led to significant changes in the microstructure, as well as the phase composition. Furnace treatment of the samples resulted in shrinkage of about 1.6%, as measured by a vernier caliper on the samples before and after heat treatment. Moreover, heat treatment led to closing of internal microcracks, and merging of small globular voids into larger pores, as shown in Figure 5a. A more detailed study in Figure 5b showed a formation of polygonal crystallites within individual splats (cf. the amorphous microstructure in Figure 2b). In most splats, the crystallites consisted of δ-Al_2O_3 (dark grains), surrounded by t-ZrO_2 (observed by SEM and confirmed by the analysis of XRD patterns). Solid state crystallization took place, with various kinetics, depending on the chemical composition of each splat, resulting in formation of δ-Al_2O_3 grains with size between 0.4–2 µm.

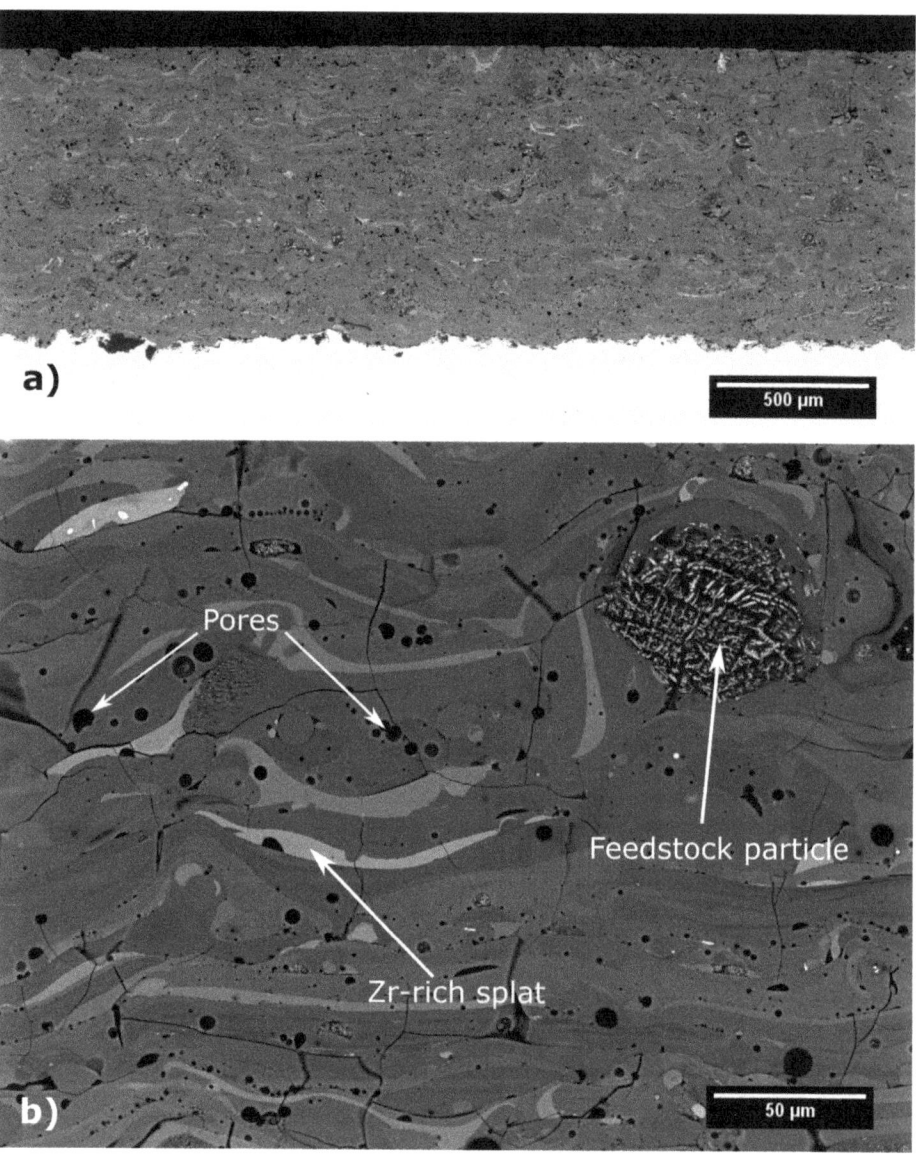

Figure 2. Microstructure of as-sprayed Al_2O_3–ZrO_2–SiO_2 coating cross-section: (**a**) coating overview; (**b**) magnified view.

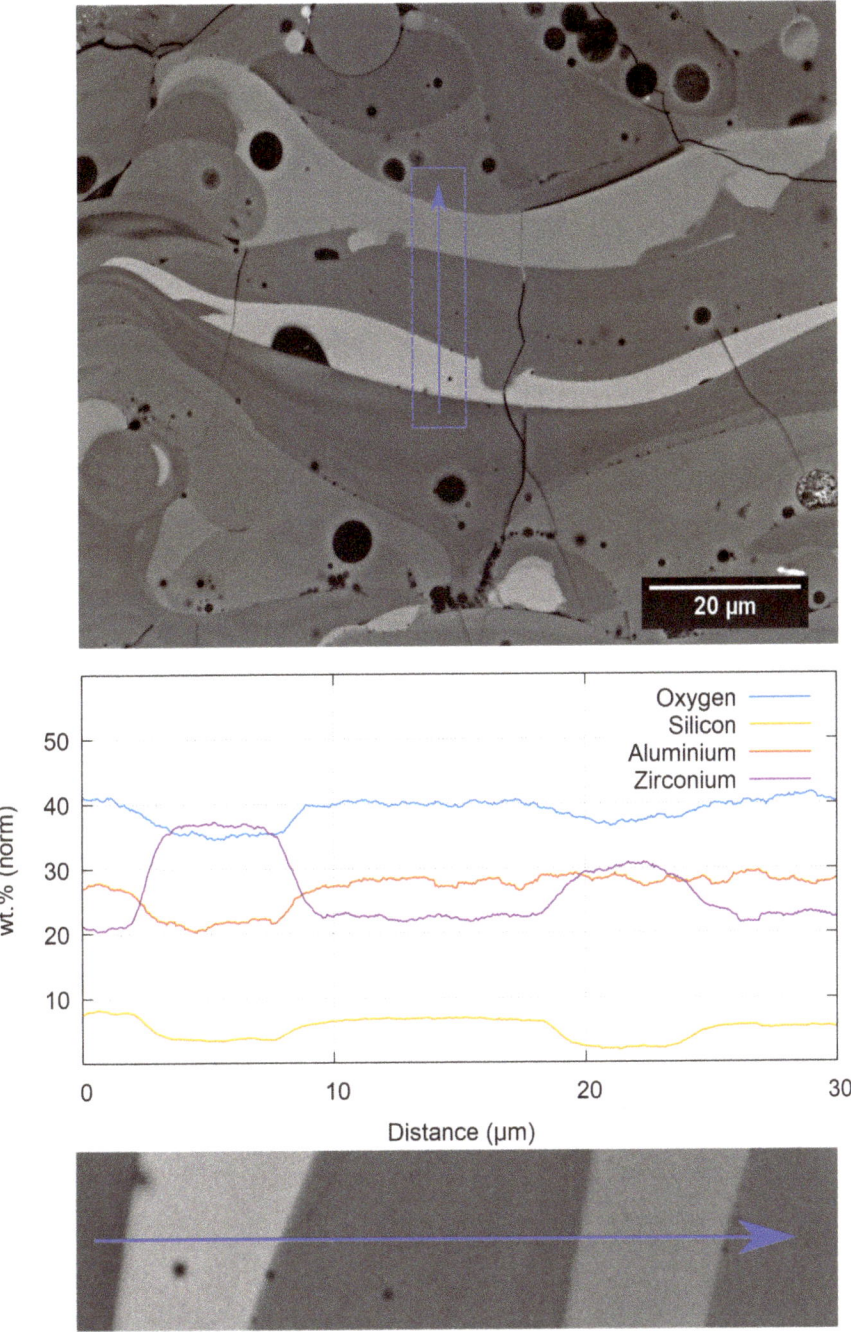

Figure 3. Local EDX line analysis of the splats in the as-sprayed material.

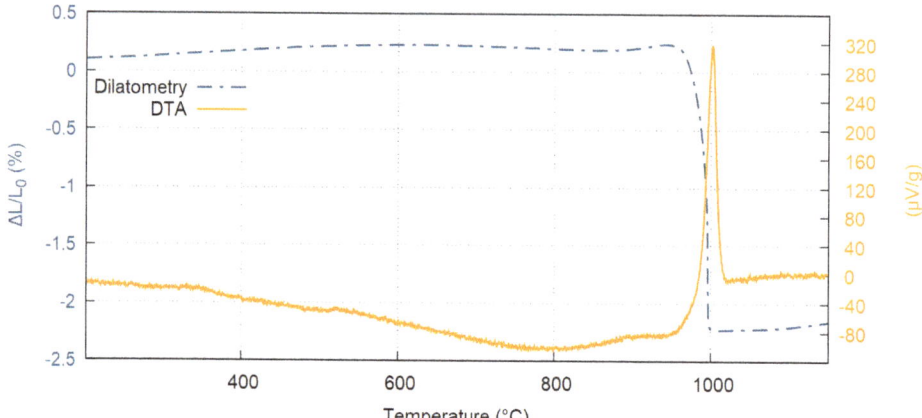

Figure 4. Measurement of differential thermal analysis (DTA) and thermal dilatometry of free standing $Al_2O_3-ZrO_2-SiO_2$ coatings.

Table 1. Used parameters of heat treatment and corresponding microhardness and wear resistance esults.

	Dwell Time (min)	Temperature (°C)	HV0.3	Cracks [1]	K (mm³/N·m)
AS	–	–	639 ± 37	0	1.3 e-2
F1	1	1050	1156± 131	0	2.9 e-4
F2	5	1050	1035 ± 180	0	5.4 e-4
	Transverse velocity (mm/min)	Power (W)	Surface HV0.3	Cracks	K (mm³/N·m)
L1	50	250	632 ± 45	0	–
L2	50	300	771 ± 89	1	1.2 e-2
L3	50	350	906 ± 87	2	
L4	50	400	939 ± 85	2	
L5	200	250	801 ± 86	0	
L6	200	300	715 ± 107	0	
L7	200	350	729 ± 146	0	1.3 e-2
L8	200	400	776 ± 143	1	1.1 e-2
L9	200	450	819 ± 143	2	–
L10	200	500	869 ± 35	3	–
L11	800	500	671 ± 146	0	–
L12	800	600	688 ± 12	0	
L13	800	800	846 ± 35	0	1.2 e-2
L14	800	1100	815 ± 22	1	7.9 e-3
L15	800	1300	863 ± 50	1	2.0 e-3
P1	3000	100,000	873 ± 21	0	1.2 e-2
P2	3000	150,000	1129 ± 169	3	–
P3	6000	100,000	645 ± 132	0	–
P4	6000	150,000	881 ± 81	1	1.1 e-2
P5	12,000	100,000	667 ± 47	0	–
P6	12,000	150,000	889 ± 69	0	1.2 e-2

0—no cracks, 1—short vertical cracks, 2—long vertical cracks, 3—vertical and horizontal cracks.

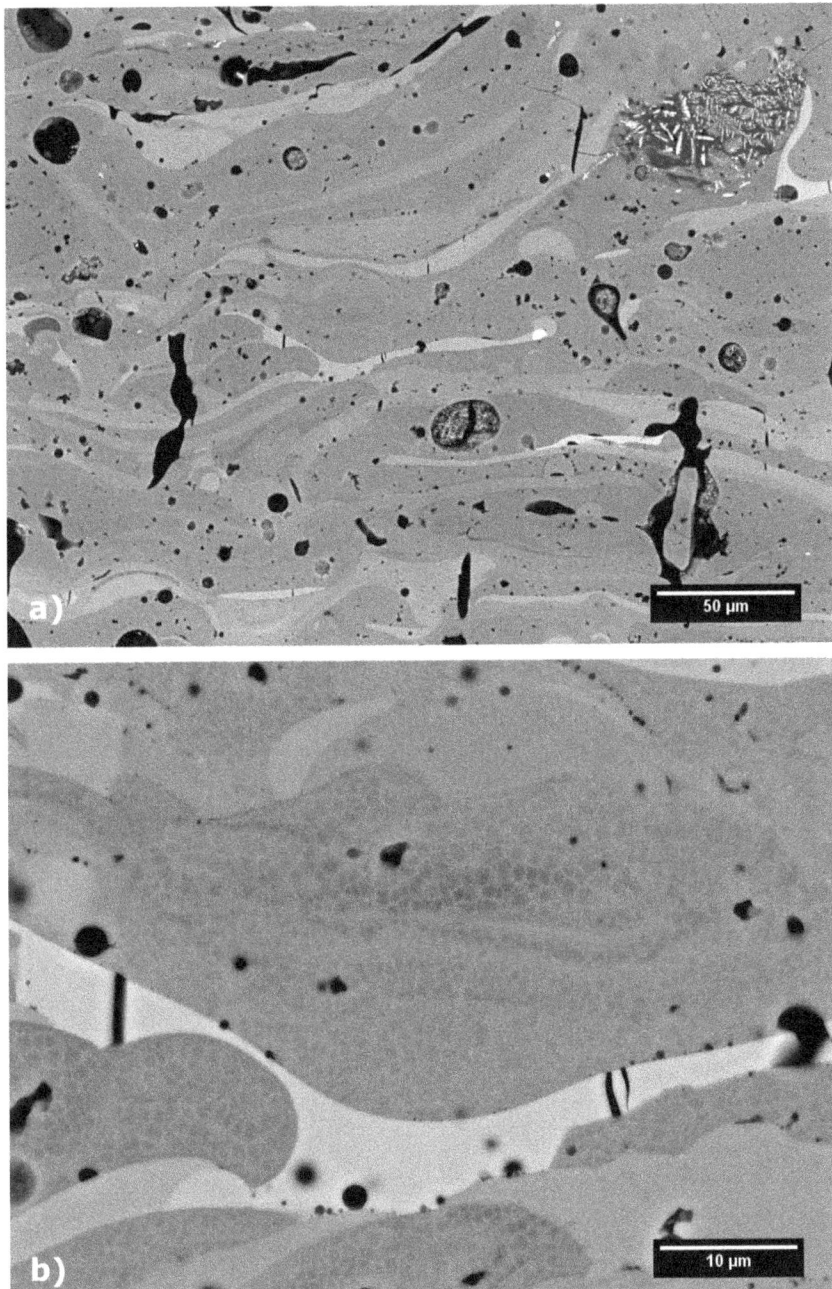

Figure 5. Cross-section of furnace heat-treated sample at 1050 °C and 5 min dwell (**a**); magnified view (**b**).

Surface heat treatment by laser and plasma also resulted in the coating crystallization. The degree of crystallization and microstructural changes in the sample depended on treatment parameters. For some samples, both laser and plasma treatment resulted in cracking of the coating. Based on this, the samples were categorized into four groups, depending on the morphology of newly developed cracks as follows: 0—no new cracks present, 1—short (<200 µm) vertical cracks, 2—long vertical cracks and 3—long vertical cracks together with horizontal cracks, triggering coating delamination. The crack classification of individual coatings is presented in Table 1. A morphology of a typical crack denoted as type 3 is depicted in Figure 6.

Figure 6. Cross-section of plasma surface heat treated sample P2 containing a major vertical and horizontal cracks, classified as type 3 in this paper. Such cracking yields the procedure unusable for applications.

3.4. Mechanical Properties

To quantify the effect of thermal treatment, a microhardness of the coating was measured. The lowest microhardness of 639 ± 37 HV0.3 was measured for the as-sprayed sample, while the highest hardness was obtained for the furnace treated sample F1 with the average value 1156 ± 131 HV0.3. In case of surface treated samples, gradually changing values of microhardness were observed, with the highest hardness measured close to the coatings free surface and lowest hardness close to the substrate. An example of such microhardness profile is provided in Figure 7 for sample L4. The depth of the influenced layer varied significantly with the heat treatment parameters from a few tens of micrometers up to 800 µm (sample P2). To facilitate a mutual comparison of all coatings, the microhardness, measured closest to the coating free surface (corresponding to the depths of approximately 87 ± 5 µm below the surface, indicated by the red cross in Figure 7), was selected as a reference value. The average values of microhardness, calculated from five indents, are listed in Table 1.

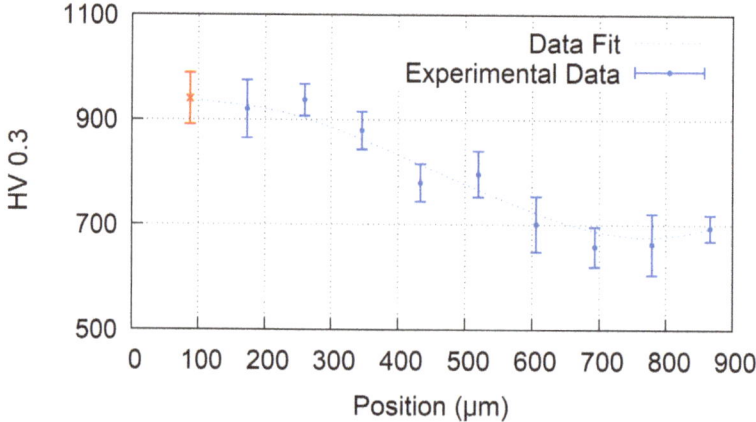

Figure 7. Microhardness profile of the laser heat treated sample L4. For samples with gradient hardness, the value closest to the surface (red color) was taken as reference.

Pin on Disc wear resistance was evaluated for selected samples. The criteria for samples selection were based on following parameters: (i) significant change in surface microhardness and (ii) cracks in the coating of type 0 or 1 only. Based on these criteria, nine surface heat treated samples were selected, along with the as-sprayed sample and two furnace heat treated samples as the reference.

The determined values of wear resistances of the coatings, presented by material volume loss K, are listed in Table 1. Graphical interpretation of hardness and wear resistance of the samples is pictured in Figure 8.

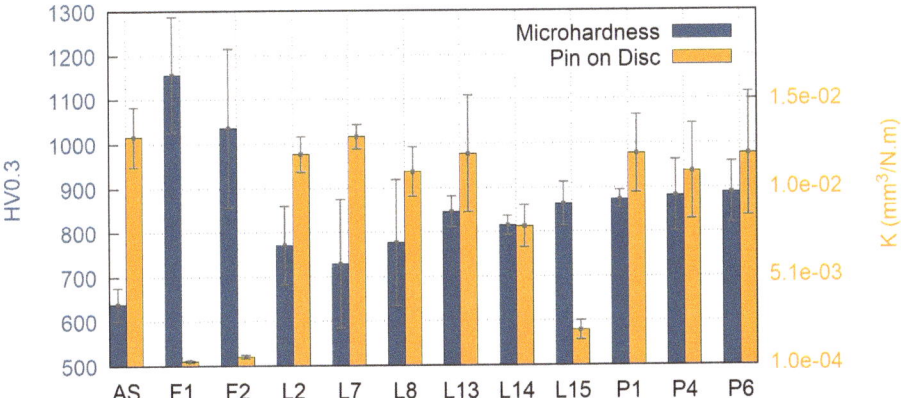

Figure 8. Microhardness and Pin on Disc volume loss for as-sprayed, furnace treated, laser treated and plasma treated samples.

4. Discussion

The original crystalline feedstock powder was transformed during spraying to almost fully amorphous coatings, as can be seen from the diffraction patterns in Figure 9 and Table 2, where FS and AS stand for feedstock powder and as-sprayed coating, respectively. Subsequently, the amorphous phase was partially transformed back to crystalline during heat treatment, forming mainly t-ZrO_2 nano-crystals, together with δ-Al_2O_3 and m-ZrO_2. Silicon dioxide, present in the original feedstock, remained in the amorphous phase, or transformed to mullite ($3Al_2O_3 \cdot 2SiO_2$), depending on the heat treatment conditions of the samples.

The furnace heat treatment resulted in almost fully crystalline samples (87% and 100% crystallinity for samples F1 and F2, respectively). As expected, both of these samples showed the highest Vickers microhardness values 1156 ± 131 HV0.3 for sample F1 and 1035 ± 180 HV0.3 for sample F2, as well as the best wear resistant behaviour with material volume loss of 2.9×10^{-4} mm^3/N·m for sample F1 and 5.4×10^{-4} mm^3/N·m for sample F2 (see Table 1 for wear rate results of all samples). This was due to the fact that, during the crystallization, the coatings could freely undergo unconstrained shrinkage, since they were removed from the substrate prior the heat treatment. Consequently, there was no CTE mismatch between the substrate and the coating, resulting in no additional cracking. Furthermore, some micro-cracks originally present in the as-sprayed material closed up by the sintering effect during the heat treatment, which improved the mechanical properties as well. However, the highest contribution to the observed increase in microhardness and wear-rate resistance may be attributed to formation of nanocrystallites of various phases within the microstructure. In particular, a formation of t-ZrO_2 is believed to have a significant influence on the improvement of mechanical properties. In order to conceive fine differences in the microstructure, crystallite size (or coherently diffracting domains' size) of the t-ZrO_2 phase was determined by the Rietveld refinement method for XRD diffractograms (see Table 2). Crystallite size of the t-ZrO_2 phase of the sample F1 showed the smallest crystallite size (14 nm) from all the measured samples.

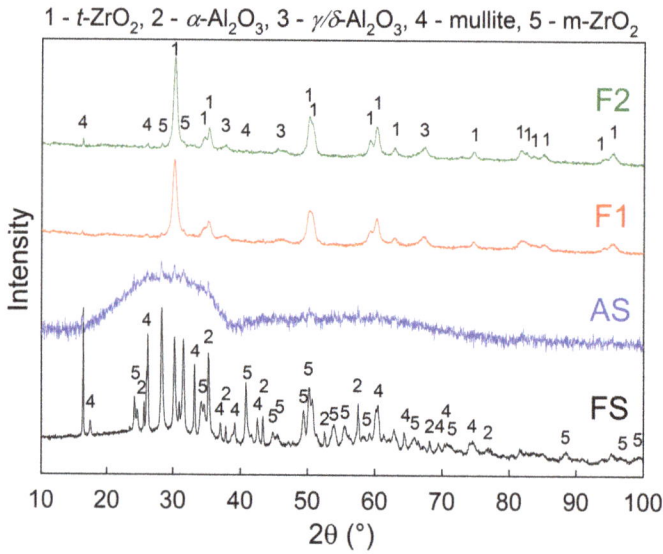

Figure 9. X-ray diffraction patterns of the feedstock (FS), as sprayed coating (AS), and furnace heat treated samples (F1 and F2).

Laser heat treated samples showed significant formation of vertical cracks, originating from the constrained shrinkage during the crystallization. Therefore, samples with no cracks or short vertical cracks only were selected for further tests, since extensive cracking of the samples may compromise its overall mechanical properties, corrosion and chemical resistance. The best mechanical properties were measured for the samples L14 and L15, which exhibited rather low wear rates of 7.9×10^{-3} mm^3/N·m and 2.0×10^{-3} mm^3/N·m, respectively. For these samples, the highest transverse velocity of the laser of 800 mm/min was used, combined with the highest laser powers of 1100 W and 1300 W, respectively. These two samples showed only limited cracking, significant increase in Vickers microhardness up to 863 ± 50 HV0.3 for sample L15 and improvement in wear resistant properties (compared to the as-sprayed coating). In fact, the sample L15 showed the best wear resistance from all surface heat treated samples. Similarly to the furnace treated samples, the sample L15 transformed to fully crystalline material, in the vicinity of the coatings surface, with the average CDD of t-ZrO$_2$ of 32 nm. The change in the microstructure and the appearance of Vickers indents are presented in Figure 10a,b. From Figure 10, formation of bright domains within individual splats was observed in sample L15. These may be segregated domains of ZrO$_2$ phase; however, they are too small for an accurate identification by EDX. Interestingly, the sample L14 remained mostly amorphous, with only 17% of the sample surface crystallized, as evaluated by the Rietveld analysis of XRD patterns measured from the sample surface. The rather incomplete crystallization was caused by the fact that the combination of 800 mm/min transverse velocity and lower power of 1100 W heated up the sample's surface just a little above the crystallization temperature. Due to cooling through substrate heat transfer, the crystallization was very limited in this case, which in turn leads to inferior wear resistance, in comparison with the sample L15.

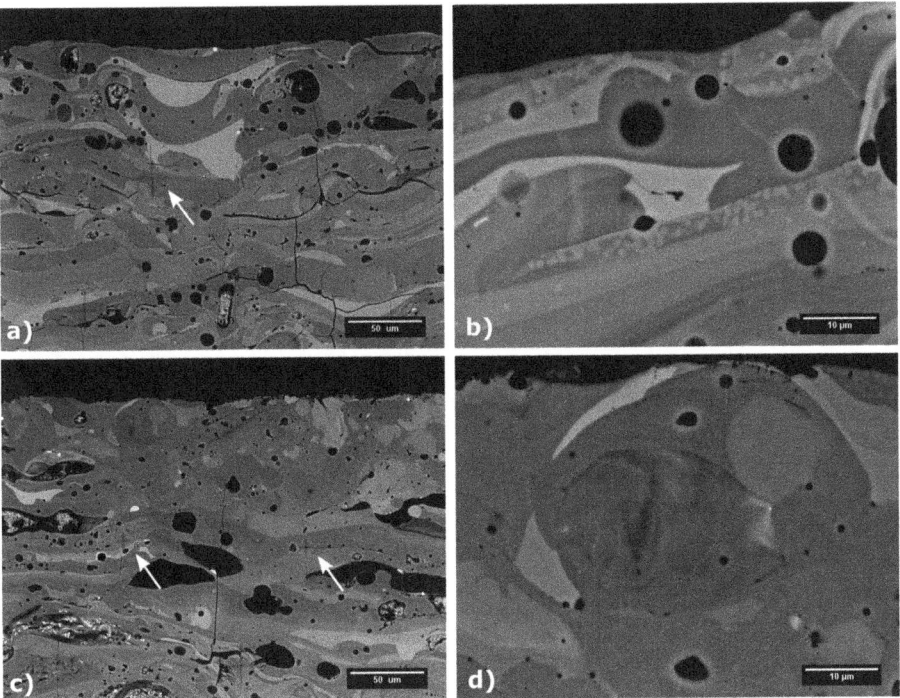

Figure 10. Comparison of the samples heat treated by laser—sample L15 (**a,b**) and plasma—sample P2 (**c,d**). Vickers indents marked by the arrows.

Table 2. Crystallinity of the samples. Ratio of t-ZrO$_2$ and its CDD size.

	Amorphous (%)	Crystalline (%)	t-ZrO$_2$ (%)	CDD (nm)
FS	16	84	10	40
AS	92	8	5	–
F1	13	87	33	14
F2	0	100	35	16
L14	83	17	29	21
L15	0	100	46	32
P2	0	100	41	23
P4	92	8	5	–

Plasma heat treatment resulted in an increase of the Vickers microhardness for the samples P1, P2, P4 and P6. Surprisingly, the related improvement of wear resistance of plasma heat treated samples was not so pronounced, compared to the as-sprayed sample. This was probably caused by the used high transverse velocity selected to prevent melting of the samples' surfaces. The high plasma torch movement speeds resulted in very short time of treatment, during which the samples were exposed to the temperatures above the crystallization point. Consequently, the grain nucleation and diffusion growth processes were limited and possibly happened only on the very specimen surface. The XRD patterns of all but one of the plasma treated samples showed only minor changes, compared to the as-sprayed samples. The only difference was the sample P2, which was produced using the highest plasma power of 150 kW and lowest transverse velocity of 3000 mm/min and its XRD pattern suggest 100% surface crystallinity, comparable to the laser treated sample L15 (see the Figure 11). However, the wear resistance properties were not measured for the P2 sample, since the most excessive cracking was observed in the SEM (see the Figure 6). In between the cracks, the cross-section of the sample P2 showed a microstructure (Figure 10) similar to the one of furnace treated samples F1 and F2. Formation of dark domains of alumina, enclosed by lighter regions, rich in ZrO$_2$ were observed in back scattered electron mode in SEM and such element distribution was confirmed by local EDX analysis. The remaining plasma heat treated samples didn't show any changes in the phase composition (compared to the as-sprayed state) and, therefore, no microstructure changes were observed in SEM for them.

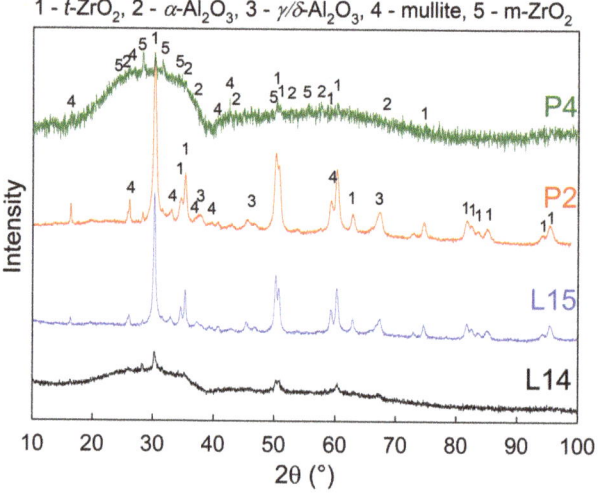

Figure 11. X-ray diffraction patterns of the laser and plasma heat treated samples.

Plasma surface heat treatment of the ceramic coating is very challenging. The power density of high enthalpy plasma torch, combined with low thermal conductivity of the ceramic coatings needs precise adjustment of heat treating conditions to provide sufficient heat treatment of the coating and, at the same time, prevent the coating from undesirable overheating (or even remelting). Therefore, further optimization of plasma heat treatment conditions, e.g., a change in the stand-off distance, or the use of multiple short passes of the plasma torch above the coating, has the potential to result in similar improvement of mechanical properties, such as was presented for the samples heat treated by laser. The analysis of the wear tracks in the SEM showed remarkable differences in the wear mechanism. The as-sprayed sample displayed rather wide (over 3 mm) wear track, reflecting its high material removal rate in the Pin on Disc test. The observed wear mechanism was mainly debonding and cracking of loosely connected splats which were crushed by the sliding ball, leaving coarse debris in the wear track. On the other hand, the furnace-treated samples showed shallow and narrow wear track (about 1.1 mm in width) filled with fine debris originating mainly from grinding off of the sintered splats. No splat debonding was observed for the furnace-treated sample. A combination of both above-mentioned mechanisms was observed for laser and plasma treated samples, where the wear tracks were filled with mixture of fine and coarse wear debris, the former originating from grinding off of the surface, and the latter formed due to debonding and cracking of splats.

5. Conclusions

Ceramic powder of near eutectic composition from the ternary system of Al$_2$O$_3$–ZrO$_2$–SiO$_2$ was plasma sprayed onto steel substrates to create 1 mm-thick coatings. Atmospheric plasma spraying was carried out by hybrid water stabilized plasma torch WSP-H 500. The as-sprayed coatings were amorphous and their hardness and wear resistance are rather low. Unfortunately, bulk furnace treatment is not applicable for coatings on metallic substrates, since it irreversibly deteriorates the properties of the substrate and causes cracking and spallation of the coating due to CTE mismatch and substrate oxidation. Therefore, the coatings were subjected to surface heat-treatment to improve mechanical properties of the coating while maintaining good adhesion to the substrate. In addition, the stripped-off coatings were subject to furnace heat treatment to obtain reference samples. The surface heat treatment by laser or plasma torch resulted in significant changes in the coating microstructure. The surface layer of the coating transformed from amorphous to nanocrystalline structure within the splats. Some of the heat treatment conditions led to formation of vertical cracks in the coatings, which compromised their overall mechanical properties. In other cases, these changes were accompanied by improvement of the coating mechanical properties. Vickers microhardness increased from 639 ± 37 HV0.3 for the as-sprayed coating, up to 1156 ± 131 and 1129 ± 169 HV0.3 for the furnace treated and selected plasma treated samples, respectively. Wear resistance was improved more than six times, from the value of material volume loss 1.3×10^{-2} mm^3/N·m of the as-sprayed sample, down to 2.0×10^{-3} mm^3/N·m for some laser treated samples. The changes in mechanical properties of the heat treated samples were caused by the solid stage crystallization in the surface layer of the originally amorphous coatings. The samples with the highest hardness and wear resistance were fully crystalline, and had a very low size of coherently diffracting domains in the range of 14–32 nm. Utilization of surface heat treatment could be an efficient final stage of coating manufacturing, since only a thin surface layer of the coating can be treated to meet the specifications for targeted wear-resistant application. Heat treatment of Al_2O_3–ZrO_2–SiO_2 coatings is a very stochastic process as each splat has different chemical composition. Therefore, solid state crystallization has different kinetics among splats, giving rise to various phases and crystalline grain sizes. Improvement of mechanical properties is then controlled mainly by formation of crystallites with sizes in tens of nanometers. Surface treatment of ceramic coatings by laser and plasma, presented in this study, was successfully used for inducing such nanocrystalline microstructure in originally amorphous material and these two methods may find application for similar materials, which tend to form amorphous coatings.

Author Contributions: J.M. is the primary study author who was involved in all experimental tasks, analysed the results and compiled the manuscript. F.L. performed XRD measurement and provided its in-depth analysis. S.C. performed measurement of dilatometry and differential thermal analysis. S.H. performed and evaluated all wear resistance tests. M.B. suggested and supervised laser treatment methodology. T.T. participated in coating deposition, metallographic samples preparation and evaluated the wear mechanisms. J.C. critically reviewed and edited the manuscript. R.M. and O.K. reviewed the manuscript in their particular areas of expertise. T.C. proposed the original idea, supervised the project, and provided funding acquisition.

Funding: The work was supported by the Ministry of Industry and Trade of the Czech Republic under project FV30058 Development of the ball valve "Top entry—KK8TE".

Acknowledgments: The authors would like to express special thanks to Dipl.-Phys. Marko Seifert for laser treatment of the samples.

Conflicts of Interest: The authors declare no conflict of interest.

References

1. Chraska, T.; King, A.H. Transmission electron microscopy study of rapid solidification of plasma sprayed zirconia—Part I. First, splat solidification. *Thin Solid Films* **2001**, *397*, 30–39. [CrossRef]
2. Ctibor, P.; Nevrla, B.; Pala, Z.; Sedlacek, J.; Soumar, J.; Kubatik, T.; Neufuss, K.; Vilemova, M.; Medricky, J. Study on the plasma sprayed amorphous diopside and annealed fine-grained crystalline diopside. *Ceram. Int.* **2015**, *41*, 10578–10586. [CrossRef]
3. Yang, K.; Rong, J.; Feng, J.; Zhuang, Y.; Zhao, H.; Wang, L.; Ni, J.; Tao, S.; Shao, F.; Ding, C. Excellent wear resistance of plasma-sprayed amorphous Al_2O_3–$Y_3Al_5O_{12}$ ceramic coating. *Surf. Coat. Technol.* **2017**, *326*, 96–102. [CrossRef]
4. Bolelli, G.; Cannillo, V.; Lusvarghi, L.; Manfredini, T. Wear behaviour of thermally sprayed ceramic oxide coatings. *Wear* **2006**, *261*, 1298–1315. [CrossRef]

5. Kiilakoski, J.; Musalek, R.; Lukac, F.; Koivuluoto, H.; Vuoristo, P. Evaluating the toughness of APS and HVOF-sprayed Al_2O_3–ZrO^2 -coatings by in-situ- and macroscopic bending. *J. Eur. Ceram. Soc.* **2018**, *38*, 1908–1918. [CrossRef]
6. Tjong, S.; Chen, H. Nanocrystalline materials and coatings. *Mater. Sci. Eng.* **2004**, *45*, 1–88. [CrossRef]
7. Koch, C.C.; Ovid'Ko, I.A.; Seal, S.; Veprek, S. *Structural Nanocrystalline Materials*; Cambridge University: Cambridge, UK, 2007.
8. Tellkamp, V.; Lau, M.; Fabel, A.; Lavernia, E. Thermal spraying of nanocrystalline inconel 718. *Nanostruct. Mater.* **1997**, *9*, 489–492. [CrossRef]
9. Tesar, T.; Musalek, R.; Medricky, J.; Kotlan, J.; Lukac, F.; Pala, Z.; Ctibor, P.; Chraska, T.; Houdkova, S.; Rimal, V.; et al. Development of suspension plasma sprayed alumina coatings with high enthalpy plasma torch. *Surf. Coat. Technol.* **2017**, *325*, 277–288. [CrossRef]
10. Chen, H.; Zhang, Y.; Ding, C. Tribological properties of nanostructured zirconia coatings deposited by plasma spraying. *Wear* **2002**, *253*, 885–893. [CrossRef]
11. Nass, R.; Campbell, R.; Dellwo, U.; Schuster, F.; Tenegal, F.; Kallio, M.; Lintunen, P.; Salatra, O.; Remškar, M.; Zumer, M.; et al. *Industrial Application of Nanomaterials—Chances and Risks. Technology Analysis*; Technical Report; Future Technologies Division of VDI Technologiezentrum GmbH: Dusseldorf, Germany, 2004.
12. Joshi, S.V.; Sivakumar, G. Hybrid Processing with Powders and Solutions: A Novel Approach to Deposit Composite Coatings. *J. Therm. Spray Technol.* **2015**, *24*, 1166–1186. [CrossRef]
13. Moign, A.; Vardelle, A.; Themelis, N.; Legoux, J. Life cycle assessment of using powder and liquid precursors in plasma spraying: The case of yttria-stabilized zirconia. *Surf. Coat. Technol.* **2010**, *205*, 668–673. [CrossRef]
14. Shek, C.H.; Wang, W.H.; Dong, C. Bulk metallic glasses. *Mater. Sci. Eng.* **2004**, *44*, 45–89.
15. Pawlowski, L.; Morgiel, J.; Czeppe, T. Amorphisation and crystallisation of phases in plasma sprayed Al_2O_3 and ZrO_2 based ceramics. *Arch. Metall. Mater.* **2007**, *52*, 635–639.
16. Chraska, T.; Neufuss, K.; Dubsky, J.; Ctibor, P.; Klementova, M. Fabrication of bulk nanocrystalline ceramic materials. *J. Therm. Spray Technol.* **2008**, *17*, 872–877. [CrossRef]
17. Tarasi, F.; Medraj, M.; Dolatabadi, A.; Berghaus, J.O.; Moreau, C. Enhancement of amorphous phase formation in alumina–YSZ coatings deposited by suspension plasma spray process. *Surf. Coat. Technol.* **2013**, *220*, 191–198. [CrossRef]
18. Pejchal, V.; Fornabaio, M.; Žagar, G.; Riesen, G.; Martin, R.G.; Medřický, J.; Chráska, T.; Mortensen, A. Meridian crack test strength of plasma-sprayed amorphous and nanocrystalline ceramic microparticles. *Acta Mater.* **2018**, *145*, 278–289. [CrossRef]
19. Chráska, T.; Hostomský, J.; Klementová, M.; Dubský, J. Crystallization kinetics of amorphous alumina-zirconia-silica ceramics. *J. Eur. Ceram. Soc.* **2009**, *29*, 3159–3165. [CrossRef]
20. Chraska, T.; Medricky, J.; Musalek, R.; Vilemova, M.; Pala, Z.; Cinert, J. Post-treatment of plasma sprayed amorphous ceramic coatings by spark plasma sintering. *J. Therm. Spray Technol.* **2015**, *24*, 637–643. [CrossRef]
21. Hrabovsky, M. Thermal Plasma Generators with Water Stabilized Arc. *Open Plasma Phys. J.* **2009**, *2*, 99–104. [CrossRef]
22. Musalek, R.; Medricky, J.; Tesar, T.; Kotlan, J.; Pala, Z.; Lukac, F.; Illkova, K.; Hlina, M.; Chraska, T.; Sokolowski, P.; et al. Controlling Microstructure of Yttria-Stabilized Zirconia Prepared from Suspensions and Solutions by Plasma Spraying with High Feed Rates. *J. Therm. Spray Technol.* **2017**, *26*, 1787–1803. [CrossRef]
23. Tagliente, M.; Massaro, M. Strain-driven (002) preferred orientation of ZnO nanoparticles in ion-implanted silica. *Nucl. Instrum. Methods Phys. Res. Sect. B Beam Interact. Mater. Atoms* **2008**, *266*, 1055–1061. [CrossRef]

© 2019 by the authors. Licensee MDPI, Basel, Switzerland. This article is an open access article distributed under the terms and conditions of the Creative Commons Attribution (CC BY) license (http://creativecommons.org/licenses/by/4.0/).

Article

Thermally Sprayed Coatings: Novel Surface Engineering Strategy Towards Icephobic Solutions

Heli Koivuluoto *, Enni Hartikainen and Henna Niemelä-Anttonen

Materials Science and Environmental Engineering, Faculty of Engineering and Natural Sciences, Tampere University, 33720 Tampere, Finland; enni.hartikainen@tuni.fi (E.H.); henna.niemela-anttonen@tuni.fi (H.N.-A.)
* Correspondence: heli.koivuluoto@tuni.fi

Received: 25 February 2020; Accepted: 19 March 2020; Published: 21 March 2020

Abstract: Surface engineering promotes possibilities to develop sustainable solutions to icing challenges. Durable icephobic solutions are under high interest because the functionality of many surfaces can be limited both over time and in icing conditions. To solve this, one potential approach is to use thermally sprayed polymer or composite coatings with multifunctional properties as a novel surface design method. In thermal spraying, coating materials and structures can be tailored in order to achieve different surface properties, e.g., wetting performance, roughness and protection against several weathering and wearing conditions. These, in turn, are beneficial for excellent icephobic performance and surface durability. The icephobicity of several different surfaces are tested in our icing wind tunnel (IWiT). Here, mixed-glaze ice is accreted from supercooled water droplets and the ice adhesion is measured using a centrifugal adhesion tester (CAT). The present study focuses on the icephobicity of thermally sprayed coatings. In addition, surface-related properties are evaluated in order to illustrate the correlation between the icephobic performance and the surface properties of differently tailored thermally sprayed coatings as well as compared those to other coatings and surfaces.

Keywords: thermal spraying; polymer coatings; flame spraying; icephobicity; ice adhesion; wettability; coating design

---

1. Introduction

Thermal spraying is used in various application fields for the production of protective coatings. In this technology, almost all materials, e.g., metals, metal alloys, ceramics, hard metals, composites and polymers, can be used as a coating material as well as a substrate or base materials. Thermal spraying consists of different spray techniques such as flame, arc, high-velocity flame, plasma and cold gas dynamic spray processes [1]. The basic idea is the same in these different processes: coating material is melted or accelerated, sprayed on the surface, solidifying or deforming, and this way building-up a coating. Thermal spray processes can use thermal (combustion or electric arc) or kinetic energy (high velocity) or a combination of these for the coating formation [1,2]. Furthermore, this is a fast and robust coating manufacturing technology and suitable for many uses. For example, one interesting application field is to produce polymer coatings by thermal spraying. This way, solvent-based coating methods can be avoided in polymer coating production, acting as an environmentally friendly coating processing method. Polymer coatings are used, e.g., to increase corrosion, wear, and environmental resistance, and as slippery surfaces for reduction of friction [3–5]. They have shown great potential to have good icephobic properties, i.e., low ice adhesion, suitable wettability and freeze-thaw performance, which makes it easier to remove accreted ice on the surface. This is under high interest because surface engineering including thermal spraying could provide a sustainable approach for icing issues. In addition, the durability of current anti-icing solutions against environmental stresses and performance in all icing conditions is insufficient and thus, novel solutions are welcome.

Thermal spray coating solutions are under development to tackle the icing issues. These inflict serious problems for various industrial operations such as offshore industry, transport and cargo, ship industry, renewable energy production and aviation [6–11]. Ice accretion on the surfaces significantly reduces the efficiency, the safety and the operational tempo of different industrial processes. These detrimental icing events take place worldwide, e.g., in Scandinavia, Europe, Russia, Northern America, Japan and China [10,12]. The on-going climate change has also decreased sea ice coverage in the Arctic Ocean, which has considerably grown the industrial activity in this area [7]. The most typical examples of icing problems can be associated to the icing of superstructures of sea vessels and offshore platforms [6], ice accretion on wind turbines blades [13] and airplane wings [9] as well as ice loads on power network structures [10] and tall structures [11]. At the worst, icing is causing disasters, which have deep socioeconomic impacts, being hazardous not only for the environment but also for personnel.

It is important to find reliable solutions for these icing challenges. Active de-icing methods have been used to remove the ice, e.g., by using heaters, which can be produced with thermal spraying of metallic materials [14,15]. We have focused on passive anti-icing methods such as icephobic surfaces. The main idea of icephobic surfaces is to reduce the ice adhesion on the surface and prevent ice accumulation on the surface [16,17]. Surface engineering and coating technologies have shown potential results, but still more development is needed especially for durable coatings and surfaces, which are not losing their icephobicity under other environmental stresses such as rain, UV light, sand, other impurities or temperature exchanges. Many superhydrophobic surfaces have good icephobicity [18] but they might not be as resistant and durable as needed in environmental conditions. On the other hand, one of the latest icephobic surfaces, which have been under high research and interest, are slippery liquid impregnated porous surfaces (SLIPS) [19,20]. They have shown low ice adhesion values, acting as icephobic surfaces. However, if the porous solid layer is very thin, durability might be limited. Therefore, other surface engineering solutions are needed. Coating design can be varied in thermal spraying. Dense or porous coatings can be produced, depending on the requirements and needs of the surface and structure. In this study, we are producing smooth flame-sprayed polymer coatings, which have low ice adhesion as well as surfaces with even lower ice adhesion by combining thermal spraying and SLIPS strategy. Porous polymer coatings can be produced by using flame spraying and then, impregnated with the lubricant. Both coating design strategies have shown their suitability for icephobic purposes [21–23].

Generally, polymer materials have high chemical and environmental resistance [24]. Especially hydrophobic polymers have also potential to act as icephobic and slippery surfaces because they possess minimal water interaction and absorption. Extending this to the coatings, smooth and dense polymer coatings can be produced by thermal spraying [3,21,22]. Actually, flame spraying is one of the common thermal spray methods for spraying polymer coatings [3,5]. Process parameters especially temperature influence on coating formation. Donadei et al. [22] have noticed that lower process temperature by using high transverse speed and higher working distance will cause less polymer degradation during spraying, which is beneficial for icephobic behavior of the coatings. The advantages of thermally sprayed polymer coatings are related to the low cost and high performance of the coatings [3,25]. Flame-sprayed polyethylene (PE) coatings have previously been studied by Vuoristo et al. [26], where the research focused on the use of flame-sprayed PE coatings as natural gas pipeline coatings. On the other hand, ultra-high molecular weight polyethylene (UHMWPE) is known to have good protective properties and flame-sprayed UHMWPE coatings have also been studied [27]. Flame-sprayed PE coatings have primary applications in corrosion protection of components and metal structures as the alternatives for paints and metallic coatings. One benefit is also their applicability in difficult processing conditions [5].

In this study, we produce polymer coating by using thermal spraying and evaluate them based on their icephobic performance. Investigations are focusing on the icephobicity and wettability of thermally sprayed polymeric coatings and several reference materials and surfaces. In the icing tests, ice is accreted in an icing wind tunnel (IWiT) and ice adhesion measured with a centrifugal ice adhesion tester (CAT). One interesting focus point was the durable icephobic slippery liquid infused porous surfaces (SLIPS) designed and manufactured by using flame spraying (FS), utilizing polymeric materials, and further, impregnated with oil. These surfaces had low ice adhesions, which express their high icephobicity.

2. Materials and Methods

2.1. Materials

Thermally sprayed icephobic surfaces were produced by using the flame spray (FS) process. Dense FS coatings were sprayed with relatively low gas flow rates in order to prevent overheating and burning of the powder with the flame. After spraying, a few post-heating passes with the flame were done without feeding powder to densify the structure and smoothen the surface. Commercially available and thermally sprayable polyethylene (PE, melting point 128 °C), PE mixed with fluoropolymer perfluoroethylene propylene (FEP, melting point 260–290 °C), ultra-high molecular weight polyethylene (UHMWPE, melting point 136 °C), polyether ether ketone (PEEK, melting point 343 °C) and polypropylene (PP, 160 °C) powders were used in these experiments. More information about flame spraying of PE, PE + FEB and UHMWPE coatings can be found from our previous study [21]. Furthermore, slippery liquid infused porous surfaces (SLIPS), combining porous FS PE coating with a lubricant was studied [23]. All FS coatings were produced using an oxygen–acetylene flame spray gun (Castodyn DS 8000, Castolin Eutectic, Lausanne, Switzerland). Gas pressure was 400 kPa for oxygen and 70 kPa for acetylene. Powder feeder (4MP, Oerlikon Metco, Pfäffikon, Switzerland) was used with compressed air as the carrier gas. Gas flows and spray distances were varied low to very low in order to achieve a porous FS PE structure. Rough PE porous surface was achieved with as-sprayed conditions whereas post-heating was done for smooth porous PE structure. After coating production, silicone oil (50 cSt, Sigma-Aldrich, Merck KGaA, Darmstadt, Germany) was impregnated into the structure in order to form the FS-SLIPS.

Other materials and coatings were studied for comparison. Bulk metals were mirror-polished aluminum (YH75, Hakudo Co., Tokyo Japan) and stainless steel (EN 1.4301/2B). Commercial paints tested included BladeRep (Alexit®, Mankiewicz Gebr. & Co., Hamburg, Germany), wind turbine paint (Carboline Ltd., St. Louis, MS, USA), Nanomyte®(NEI Corporation, Somerset, NJ, USA) and NeverWet®(NeverWet LLC, Lancaster, PA, USA). UltraEverDry®film (NetDesign s.r.o, Liberec, The Czech Republic) was representing a superhydrophobic surface. Bulk polymers were polyethylene (PEHWU, Simona AG, Kirn, Germany), polypropylene (PP-DWU AlphaPlus, Simona AG, Kirn, Germany), polytetrafluoroethylene (PTFE G400, Guarniflon, Castelli Calepio, Italy) and ultra-high molecular weight polyethylene (UHMW-PE, Tivar®1000, Quadrant Group, Zurich, Switzerland). Other references were PTFE tape (PTFE Extruded Film Tape, 5490, 3M™, St. Paul, MN, USA), low-density polyethylene sheet (LDPE, thick film used for paper making [28]) and SLIPS containing thin polymeric membrane and infused silicone oil. PTFE and PP membranes (0.2 µm pore size, Sterlitech Inc., Kent, WA, USA) were used together with silicone oil (50 cSt, Sigma-Aldrich, Merck KGaA, Darmstadt, Germany). Preparation of these SLIPS surfaces is shown in our earlier research [20]. All coatings, materials and surfaces studied are summarized in Table 1.

Table 1. Coatings and surfaces used in this study divided into different material groups.

	Surface Treatment	Material	Form
Flame-Sprayed Coatings			
FS PE [1]	[1] As-sprayed	Polyethylene	Coating
FS PE [2]	[2] Polished	Polyethylene	Coating
FS PE+FEB [1]	[1] As-sprayed	Polyethylene + perfluoroethylene propylene	Coating
FS PE+FEB [2]	[2] Polished	Polyethylene + perfluoroethylene propylene	Coating
FS UHMWPE [1]	[1] As-sprayed	Ultra-high molecular weight polyethylene	Coating
FS PEEK [1]	[1] As-sprayed	Polyether ether ketone	Coating
FS PP [1]	[1] As-sprayed	Polypropylene	Coating
FS-SLIPS_1	SLIPS	Polyethylene (fine) + silicone oil (50 cSt)	SLIPS
FS-SLIPS_2	SLIPS	Polyethylene (coarse) + silicone oil (50 cSt)	SLIPS
Bulk Polymers			
PE [2]	[2] Polished	Polyethylene	Bulk plate
UHWPE [2]	[2] Polished	Ultra-high molecular weight polyethylene	Bulk plate
PP [2]	[2] Polished	Polypropylene	Bulk plate
PTFE [2]	[2] Polished	Polytetrafluoroethylene	Bulk plate
Bulk Metals			
Al [2]	[2] Polished	Aluminum	Bulk plate
SS [2]	[2] Polished	Stainless Steel	Bulk plate
References			
LDPE	As-received	Low-density polyethylene	Film
PTFE	As-received	Polytetrafluoroethylene	Tape
SLIPS (PTFE)	SLIPS	Polytetrafluoroethylene membrane (0.2 µm) + silicone oil (50 cSt)	SLIPS
SLIPS (PP)	SLIPS	Polypropylene membrane (0.2 µm) + silicone oil (50 cSt)	SLIPS
Commercial Paints			
BladeRep9	Painted, BR	Polyurethane	Paint
Carboline	Painted, C	Elastomeric	Paint
Nanomyte	Painted, NM	Nanocomposite	Paint
NeverWet	Painted, NW	Superhydrophobic	Paint
UltraEverDry	Sprayed	Superhydrophobic	Film

[1] As-sprayed surface, [2] Polished surface

2.2. Test Methods

Structures of FS coatings were analyzed with an optical microscope (Leica DM2500 M, Wetzlar, Germany) from the cross-sectional coating samples. Wetting behavior and water contact angle measurements were done with a drop shape analyzer (DSA100, Krüss, Hamburg, Germany). Static contact angle (CA), advancing contact angle (ACA) and receding contact angle (RCA), as well as contact angle hysteresis (CAH), were studied. The experiments were achieved by dispersing 5 µL water droplets of ultra-high purity water (MilliQ, Merck KGaA, Darmstadt, Germany) onto surfaces. In addition, roll-off/sliding angles were performed for SLIPS by tilting the surfaces with a 10 µm water droplet. The roughness of the solid and dry surfaces was analyzed by an optical profilometry (Alicona Infinite Focus G5, AT, Graz, Austria) using a 20× objective magnification, achieving a vertical resolution of 50 nm. The area of the measurements was 0.81 mm × 0.81 mm in the XY-plane and the results are presented as surface roughness, Sa, values.

Icing tests were done at Tampere University (TAU). The mixed-glaze ice type was accreted in the icing wind tunnel (IWiT) and the ice adhesion strength was examined with a centrifugal ice adhesion tester (CAT) [29]. Icing test systems are presented in Figure 1.

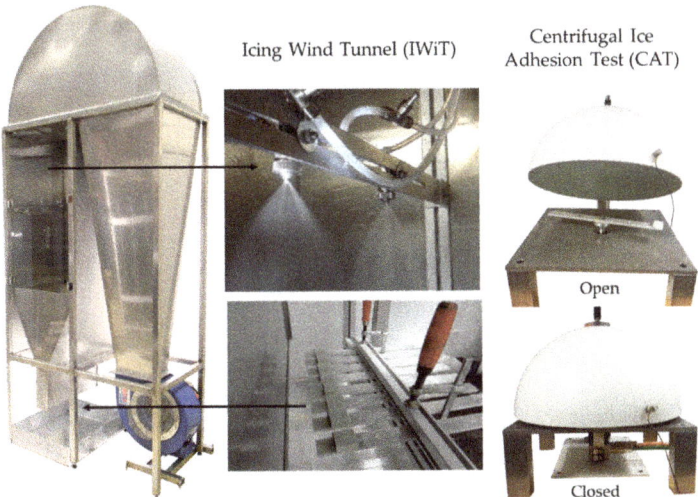

Figure 1. Icing Wind Tunnel (IWiT) and Centrifugal Ice Adhesion Test (CAT) at Tampere University.

These test facilities are located in the cold climate room, where temperature can be set from room temperature to as low as −40 °C. All tested samples were 60 mm × 30 mm in size and the ice accreted area was 30 mm × 30 mm. Laboratory grade II+ water (Purelab Option-R 7/15, Elga, UK) was used for accreting the ice. The samples were let to cool down in the cold room prior to the ice accretion process. The most important parameters for the ice accumulation procedure for mixed-glaze ice used in this study were the ambient temperature of the cold room (−10 °C), the wind speed (25 m/s), and the supercooled water droplet size (~30 µm, as given by the nozzle manufacturer). In this study, the maximum water flow rate was 0.3 L/min. The accreted glaze ice refers to ice that has characteristics from both rime and glaze ice types, having more glaze-like features without icicles or runback ice. The mixed-glaze ice type forms a clean-cut rectangular structure. An example of the accreted ice is presented in Figure 2. In addition to this mixed-glaze ice, rime and glaze ice can be accreted in the IWiT as shown in our previous studies [21,29]. Ice adhesion strengths were determined using a CAT [29]. In this test, the centrifugal force detaches the ice from the surface and the ice adhesion strength (T) is calculated according to Equation (1):

$$T = F/A = mr(\alpha t)2/A \qquad (1)$$

where a piece of ice of known mass *m* and contact area *A* is spun along a radial length *r*, which spins with a constant angular acceleration α of 300 rpm/s. From this equation, the ice adhesion strength (T) via shear stress is calculated at the time of detachment *t*.

Figure 2. Mixed-glaze ice accreted in IWiT for CAT testing.

3. Results and Discussion

Thermal spraying is a versatile coating production technique. Traditionally, it is used, e.g., for corrosion and wear protection, thermal conductivity and insulation [3,30–33]. Lately, thermally sprayed coatings are developed for other advanced purposes, e.g., for self-cleaning and anti-icing [21,22,32]. Coatings and surfaces with good icephobic properties are potential to use in anti-icing applications, e.g., in wind energy, transportation, aviation and building industries. In our previous studies, flame spraying was used to produce polymer coatings with icephobic properties [20,21]. There are two approaches to achieve these goals. The first coating design approach is to produce dense and smooth polymer coatings and the second is to produce polymer coatings with porous structures and add lubricants to the structures, having slippery properties. The latter acts as SLIPS. Figure 3 shows a schematic presentation of these two flame spray approaches towards anti-icing solutions. We are focusing on thermal spraying because durable coatings with a high variety of coating thicknesses can be achieved. FS polymer coatings have shown high durability compared to paints [21] and are considered to have higher structural durability than thin SLIPS combined with a thin membrane and infused oil [20].

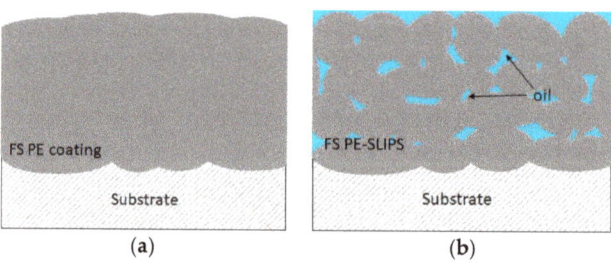

Figure 3. Schematic presentation of icephobic surfaces produced by using flame spraying (a) dense flame-sprayed (FS) polyethylene (PE) coating and (b) FS PE slippery liquid impregnated porous surfaces (SLIPS; porous structure with impregnated oil).

3.1. Coating Structures

Flame-sprayed PE, PEEK and PP polymer coatings own dense coating structures as presented in Figure 4. These coating structures correspond to the schematic presentation of dense coating in Figure 3a. Process parameters and heat-input affect the coating formation, adhesion between coating and substrate as well as denseness or porosity level inside the coating [22]. If a dense coating structure is produced, the temperature cannot be too high in order to avoid the defects, e.g., gas bubbles, caused by high-temperature gas. On the other hand, if the temperature is too low, particles are not melting enough, and pores can be formed to the structure. These dense FS polymer coatings are good examples of the icephobic coatings followed by the first approach of the coating design.

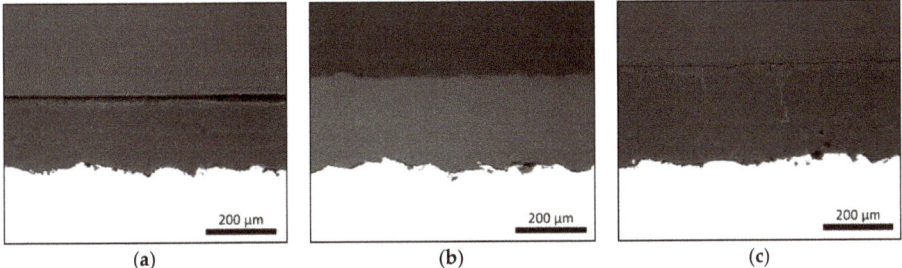

Figure 4. Flame-sprayed polymer (**a**) PE, (**b**) polyether ether ketone (PEEK) and (**c**) polypropylene (PP) coatings (middle part) on steel substrates (white part). Cross-sections. OM images.

The second coating design approach relies on the SLIPS strategy. A porous coating structure was produced by flame spraying, and afterward, lubricant was impregnated to the structure. Flame-sprayed porous PE coating impregnated with silicone oil has shown excellent icephobicity [23]. Figure 5 shows the structure of the FS porous PE coating. Open and overall porosity can be detected in the structure, which is beneficial for lubricant impregnation.

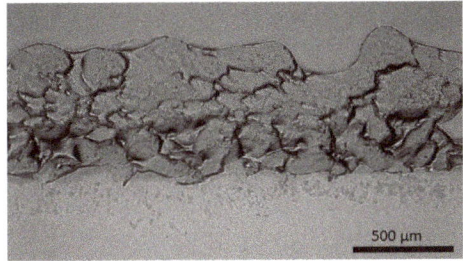

Figure 5. Structure of the FS porous polymer PE coating. Cross-section. OM image.

3.2. Wettability of the Surfaces

In this study, FS PP and PE coatings, as well as FS-SLIPS, were hydrophobic or very close to that, Figure 6. This implies that the surfaces can resist the droplet from spreading. Generally speaking, hydrophobic and superhydrophobic surfaces have advantages in self-cleaning and anti-wetting purposes. Furthermore, in this study surface state after processing affected the wettability. Microstructures and surface topographies of the plasma sprayed coatings have shown to have an effect on wettability as presented, e.g., by Xu et al. [34] and Sharifi et al. [35]. Here, the degree of coating surface roughness can vary after flame spray processing, which can be smoothened by polishing as a post-treatment. Polished FS PE coating was hydrophobic (CA 90°) whereas as-sprayed coating was slightly hydrophilic (CA 85°). Surface roughness and uniformity influenced here by changing the wettability of the coatings. Based on this, FS coatings with certain surface quality have beneficial anti-wetting properties. Donadei et al. [22] have shown small differences in the wetting behavior between FS PE coatings sprayed with different spray parameters. Heat-input was varying between spray parameters and this influenced polymer degradation, surface roughness and thus, wettability. However, all FS PE coatings were hydrophobic in that study.

In addition to contact angle (CA) values, advancing (ACA) and receding (RCA) contact angle values, as well as contact angle hysteresis (CAH), indicate the wetting performance of the surfaces. CAH can be derived from the difference between ACA and RCA, demonstrating the overall droplet mobility. High CAH indicates differences between the ACA and RCA, which results from surface properties, such as local roughness variations, surface free energy, surface chemistry leading to altering droplet behavior.

Flame-sprayed dense coatings have reasonable high contact angle hysteresis, but it can be reduced with FS-SLIPS surfaces. FS-SLIPS have low contact angle hysteresis and, therefore, water droplet mobility is high. This, in turn, is beneficial for slipperiness and the slippery properties of the surfaces.

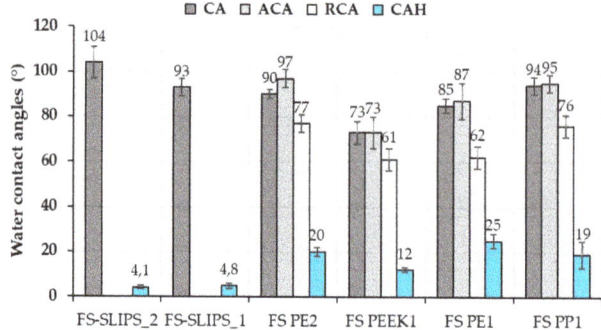

Figure 6. Water contact angles for flame-sprayed (FS) polymer coatings and FS-SLIPS. CA: static contact angle; ACA: advancing contact angle; RCA: receding contact angle; and CAH: contact angle hysteresis. Coatings are as-sprayed (1) or polished (2) prior testing.

Wettability of the different surfaces and coatings is presented in Table 2. In addition, ice adhesion values and surface roughness (Sa values) are collected to the table. In this study, different material and surface groups have been analyzed in order to get an understanding of the behavior and potential of FS polymer coatings for application areas, where icephobicity and/or non-wettability are the important properties, such as in energy, construction and building industries.

Table 2. Overview of the results divided by different material groups. Ice adhesion, contact angle (CA), advancing contact angle (ACA), receding contact angle (RCA), contact angle hysteresis (CAH) and surface roughness (Sa) of the materials and surfaces (± standard deviation).

Material or Surface	Ice Adhesion (kPa)	Contact Angle (°)	Adv. Contact Angle (°)	Rec. Contact Angle (°)	CA Hysteresis (°)	Roughness Sa (µm)
Flame-Sprayed Coatings						
FS PE [1]	69 (±9) [21]	85 (±3)	87 (±8)	62 (±5)	25 (±3)	2.05 [21]
FS PE [2]	54 (±10) [21]	90 (±2)	97 (±4)	77 (±4)	20 (±2)	0.64 [21]
FS PE + FEB [1]	79 (±7) [21]	-	-	-	-	3.95
FS PE + FEB [2]	53 (±16) [21]	75 (±5)	82 (±5)	65 (±)	18 (±0.1)	0.96 [21]
FS UHMWPE [1]	130 (±33) [21]	91 [21]	-	-	-	1.65 [21]
FS PEEK [1]	61 (±13)	73 (±5)	73 (±7)	61 (±5)	12 (±1)	-
FS PP [1]	119 (±18)	94 (±4)	95 (±4)	76 (±5)	19 (±6)	-
FS-SLIPS_1	27 (±6) [23]	93 (±4)	*	*	4.8 (±1) *	3.0 ** [23]
FS-SLIPS_2	21 (±5) [23]	104 (±7)	*	*	4.1 (±0.9) *	38.8 ** [23]
Bulk Polymers						
PE [2]	43 (±3) [36]	95 (±2)	100 (±2)	84 (±2)	16 (±0.6)	0.64
UHWPE [2]	62 (±4) [36]	84 (±2)	-	-	-	-
PP [2]	60 (±8) [36]	93 (±2)	-	-	-	-
PTFE [2]	41 (±2) [36]	101 (±0.4)	105 (±4)	78 (±6)	26 (±2)	0.38
Bulk Metals						
Aluminum [2]	343 (±35) [21]	64 (±2)	74 (±1)	56 (±3)	19 (±3)	0.26 [21]
Stainless steel [2]	269 (±13) [21]	67 (±1)	85 (±1)	63 (±3)	23 (±4)	0.23 [21]

Table 2. Cont.

Material or Surface	Ice Adhesion (kPa)	Contact Angle (°)	Adv. Contact Angle (°)	Rec. Contact Angle (°)	CA Hysteresis (°)	Roughness Sa (μm)
References						
LDPE-film	110 (±20) [28]	-	-	-	-	-
PTFE tape	39 (±14) [20]	109 (±3)	116 (±1)	108 (±3)	8 (±2)	-
SLIPS (PTFE$_{memb}$ + sil.oil)	13 (±2) [20]	98 (±2) [20]	*	*	1 (±0.5) * [20]	-
SLIPS (PP$_{memb}$ + sil.oil)	33 (±7) [20]	105 (±2) [20]	*	*	4 (±2) * [20]	-
Commercial Paints/Films						
Blade Rep9	88 (±5) [21]	84 (±2)	91 (±0.6)	62 (±0.5)	30 (±0.7)	2.24 [21]
Carboline	57 (±8) [36]	85 (±1)	89 (±3)	65 (±2)	24 (±0.4)	2.55
Nanomyte	40 (±5) [36]	101 (±1)	104 (±2)	91 (±6)	13 (±4)	8.1
NeverWet	68 (±15)	136 (±3)	141 (±1)	138 (±1)	3.3 (±0.1)	13.71
UltraEverDry	40 (±3) [29]	-	-	-	-	-

[1] As-sprayed surface; [2] polished surface; * from roll-off angle measurements; ** Sa value of porous FS PE coating without oil; - not analyzed.

Roughness of the surface has influenced the wettability [37,38] together with other surface and material properties [39]. It can be beneficial at a certain level. For SLIPS surfaces, roughness, together with porosity, can help the oil to lubricate the structure and surface. On the other hand, if a dense coating is produced by using FS, smoothness plays a role in the surface properties of the coatings. Polished surfaces are smoother compared to the as-sprayed coatings. During the thermal spray process, roughness forms due to the nature of the particle adherence and coating formation from the particles and splats. Particle and splat sizes are affecting the roughness as well as post-heating of the surface. If the surface was strongly post-heated, it resulted as a smoother surface. Especially, this is the case with thermal spraying of polymer materials because they are heat-sensitive materials, having low melting points [3].

3.3. Ice Adhesion of the Surfaces

Ice adhesion of the surfaces can be measured with different techniques and this affects strongly the given values [36]. Therefore, it is important to describe ice accretion methods and parameters together with the results. In addition to this, ice adhesion measuring technique needs to explain, e.g., is it a centrifugal, pendulum or pushing/pulling type of testing. At the moment, there is not a clear way to compare the results measured with the different tests. However, the trend of the results can be seen and compared. In this study, all the tested surfaces were measured in the same way using the same ice accretion and ice adhesion test method. Therefore, the results can be directly compared. The centrifugal ice adhesion test (CAT) used in this study has been presented as a usable test method to screen different surfaces also by Laforte et al. [16].

Ice adhesion is one of the indicators for icephobicity in such a way that low ice adhesion reflects good icephobicity and the ice is easily removed from the surface. If ice adhesion is high, then the ice is strongly adhered to the surface and icephobicity is low. We have used this definition for the ice adhesion strengths below 50 kPa, medium-low below 100 kPa, medium below 150 kPa and when high ice adhesion strengths are higher than 150 kPa. Ice adhesion is kept as extremely low when ice the value is below 10 kPa [20]. Figure 6 presents the ice adhesion strengths of tested surfaces. They were divided into different material groups in order to understand the behavior of different materials and surfaces. Metal surfaces had the highest ice adhesions due to their chemical properties whereas polymeric materials have generally relatively low ice adhesions due to their surface properties and slippery nature. From an application point of view, commercial paints were studied because they are used in the conditions where they can face the icing conditions, e.g., in wind turbine blades. The lowest ice adhesion strengths have been measured with SLIPS using porous membranes together with impregnated oil [20]. Thermally sprayed, here FS, polymer coatings showed medium-low ice adhesions. These are potential results taking account of the fact that durable coatings can be manufactured with thermal spray processing [21]. In addition, we combined durable FS PE coating and SLIPS concept and produced porous FS PE coating and impregnated oil to the porous structure [23]. This combination resulted in low ice adhesions as can be seen in Figures 7 and 8. Flame-sprayed PE and PE-based composites, as well as PEEK coatings, had medium-low ice adhesions and FS UHMWPE and PP coatings and, in turn, medium ice adhesions. However, in flame spraying, coating properties can be influenced by process parameters [22] and further improvements are possible. The durability of the traditional SLIPS can be a challenge and, therefore, this novel way to produce SLIPS with high structural durability is shown to be the potential icephobic surface engineering solution. In addition to this, many icephobic surfaces rely on small scale, expensive or multistep methods, e.g., lithography [40], chemical synthesis [41] or sol-gel coatings [42], which can be avoided with thermal spraying. This, in turn, promotes thermal spraying to be a potential solution for smart and functional coating production [43], acting as environmentally friendly surface engineering solutions.

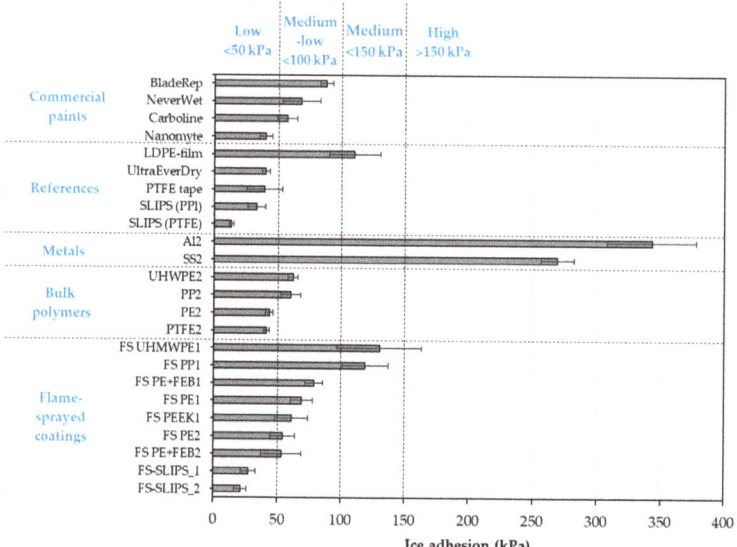

Figure 7. Ice adhesion strengths for several coatings and materials. Mixed-glaze ice accreted in IWiT and ice adhesion measured with CAT. One means as-sprayed and 2 polished surfaces. FS-SLIPS_1 contains a smoother porous structure and FS_SLIPS_2 a coarser porous structure.

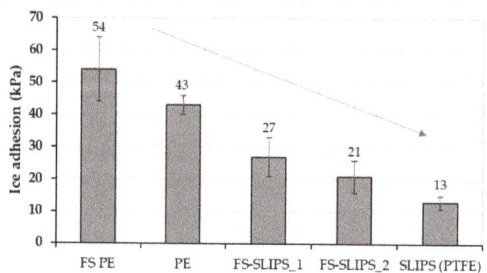

Figure 8. Comparison of ice adhesion between polished FS PE, polished bulk PE, FS-SLIPS and reference SLIPS (PTFE$_{memb\,+\,sil.oil}$).

3.4. Comparison between Icephobicity and Surface Properties

Based on our previous studies [20,22,36] and the present results, the icephobicity, in general, does not have a correlation with high water contact angles in every surface technology. Icephobicity is affected by different aspects and clear explanations of the effect of different surface properties cannot be done. However, it was shown that the wettability and wetting properties of the surfaces had an influence on the ice adhesion. In many cases, hydrophobicity is beneficial for low ice adhesion. Superhydrophobic surfaces have shown their potential for icephobicity [18] but their environmental durability is a challenge [44]. In this study, surfaces, which had low or medium-low ice adhesions, were hydrophobic whereas metal surfaces with high ice adhesions were clearly hydrophilic, Figure 9 and Table 2. Similar findings have been reported by other research as well [45]. A comparison between ice adhesions and static water contact angles (CA) is drawn in Figure 9.

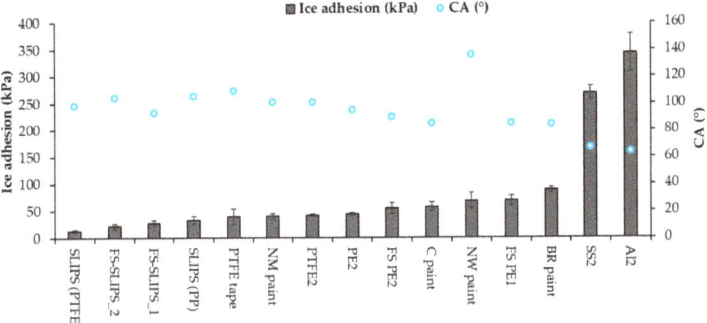

Figure 9. Ice adhesion and static contact angle relationships of the selected materials and surfaces.

Contact angles (CA) showed only local wetting behavior of the surface and better indicator could be water contact angle hysteresis (CAH) measured with dynamic contact angle or sliding/roll-off contact angle measurements. This could indicate droplet mobility and can be related to icing as well with some materials and surfaces. The CAH of the surfaces is presented in Figure 10. For SLIPS, it has been calculated from roll-off angle measurements whereas for other surfaces in this study from dynamic contact angle measurements. Therefore, also advancing and receding contact angles for other surfaces than SLIPS are presented in Figure 10. As stated, the static contact angle does not clearly explain the ice adhesion, but a better relationship can be found by comparing contact angle hysteresis [46]. This was observed also here in the case of hydrophobic surfaces. SLIPS surfaces with low CAH had the lowest ice adhesions (Figure 10).

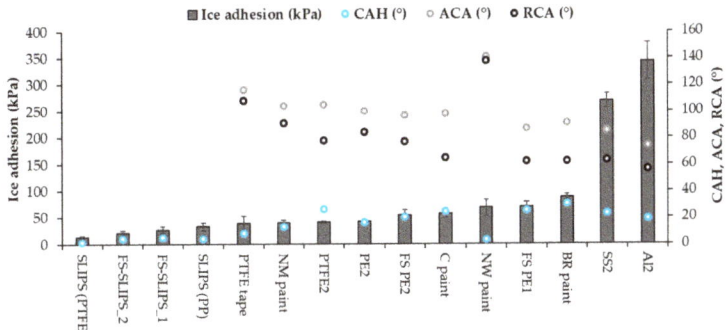

Figure 10. Ice adhesion and CAH, ACA and RCA relationships of selected materials and surfaces. CAH: contact angle hysteresis, ACA: advancing contact angle; and RCA: receding contact angle.

Roughness can influence ice adhesion although it is not a dominant factor itself. For example, mechanical interlocking as the adhesion mechanism between ice and surface can have an influence on the ice adhesion of textured surfaces [40]. Chemical properties, wettability and material properties play an important role, but it is difficult to separate their influences. Generally, higher roughness leads to surfaces with a larger number of crevices, which, in turn, gives more space for water droplets to stay. Therefore, it can increase ice adhesion. However, sometimes it is beneficial, for example, with the FS-SLIPS. FS-SLIPS_1 (fine) had lower surface roughness compared to FS-SLIPS_2 (coarse) without oil, and this resulted in lower ice adhesion for FS_SLIPS_2, when there was more area for oil to penetrate and stay on the surface. Some researchers have found that high roughness can be beneficial for lower ice adhesion [47], depending on the surface roughness pattern. Generally, it was reported that a smoother surface has lower ice adhesion [37,48,49]. This is also seen while comparing the same materials as here as-sprayed FS PE coatings and polished FS PE coatings. However, there are more dominant factors such as chemistry and surface energy and, therefore, metals (aluminum and stainless steel) had very high ice adhesion even though they are the smoothest surfaces. Figure 11 shows surface roughnesses and ice adhesions for selected surfaces. As a conclusion here, certain roughness is beneficial for low ice adhesion of the paints but with the FS polymer coatings and bulk polymers, a smoother surface gave lower ice adhesion. NW paint had high surface roughness compared to others and still reasonable ice adhesion. This could be explained by its high hydrophobicity and low CAH (Figure 10) [46].

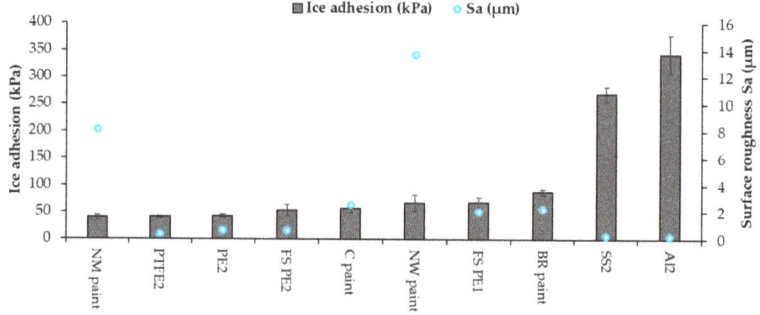

Figure 11. Ice adhesion and surface roughness (Sa) for selected materials and surfaces.

4. Conclusions

Thermal spraying has shown its potential to produce multifunctional polymer coatings for icing conditions. Flame-sprayed polymer coatings had medium-low ice adhesions, indicating their good icephobicity. Furthermore, most FS coatings were hydrophobic, which, in turn, showed their potential for anti-wettability conditions. Structurally dense and well-adhered PE, PEEK and PP coatings were produced by using flame spraying. This is advantageous for environments where protection is needed. Flame spraying is the robust and fast coating manufacturing method, which is important in several industrial applications. Furthermore, the polymers used in this study are cheap materials, showing the potential of this processing-material combination for anti-icing and anti-wetting purposes.

Furthermore, SLIPS are interesting options as icephobic solutions. The lowest ice adhesion values were achieved with traditional SLIPS as well as with FS-SLIPS as a novel surface engineering solution by combining flame spraying of porous structure with impregnation of the oil. This way, we can have low ice adhesion surfaces as SLIPS, together with a durable structure, as thermally sprayed coatings. FS-SLIPS are shown their potential to act as slippery and icephobic surfaces.

Future work will focus on the widening of material selection for icephobic thermally sprayed coatings and testing of their durability under different environmental conditions with combining icephobicity. Laboratory scale icing testing is a good way to make the ranking of the surfaces, but application-related testing is needed for further development towards specific requirements in different conditions.

Author Contributions: Conceptualization, H.K.; writing, H.K., H.N.-A.; supervision, H.K.; investigation, E.H., H.N.-A. All authors have read and agreed to the published version of the manuscript.

Funding: This research was funded by Academy of Finland, project "Thermally Sprayed slippery liquid infused porous surface – towards durable anti-icing coatings" (TS-SLIPS).

Acknowledgments: Authors would like to thank Mikko Kylmälahti of Tampere University, Thermal Spray Center Finland (TSCF), Tampere, Finland, for spraying the coatings and Valentina Donadei and Christian Stenroos of Tampere University, Tampere, Finland, for assistance in the icing laboratory at Tampere University. The authors also acknowledge "New cost/effective superhydrophobic coatings with enhanced bond strength and wear resistance for application in large wind turbine blades" (Hydrobond) project funded by EU/FP7-NMP-2012-SMALL-6.

Conflicts of Interest: The authors declare no conflicts of interest.

References

1. Pawlowski, L. *The Science and Engineering of Thermal Spray Coatings*, 2nd ed.; John Wiley & Sons Ltd.: Hoboken, NJ, USA, 2008; 626p. [CrossRef]
2. Davis, J.R. *Handbook of Thermal Spray Technology*; ASM International: Novelty, OH, USA, 2004; 338p.
3. Petrovicova, E.; Schadler, L.S. Thermal spraying of polymers. *Int. Mater. Rev.* **2002**, *47*, 169–190. [CrossRef]
4. Leivo, E.; Wilenius, T.; Kinos, T.; Vuoristo, P.; Mäntylä, T. Properties of Thermally Sprayed Fluoropolymer PVDF, ECTFE, PFA and FEP Coatings. *Prog. Org. Coat.* **2004**, *49*, 69–73. [CrossRef]
5. Winkler, R.; Bultmann, R.F.; Hartmann, S.; Jerz, A. Thermal Spraying of Polymers: Spraying Processes, Materials and New Trends. In Proceedings of the Thermal Spray, Advancing the Science and Applying the Technology, Orlando, FL, USA, 5–8 May 2003; Moreau, C., Marple, B., Eds.; ASM International: Novelty, OH, USA, 2003; pp. 1635–1638.
6. Ryerson, C. Ice protection of offshore platforms. *Cold Reg. Sci. Technol.* **2011**, *65*, 97–110. [CrossRef]
7. Moore, G. A climatology of vessel icing for the subpolar North Atlantic Ocean. *Int. J. Climatol.* **2013**, *33*, 2495–2507. [CrossRef]
8. Carriveau, R.; Edrisy, A.; Cadieux, P.; Mailloux, R. Ice Adhesion Issues in Renewable Energy Infrastructure. *J. Adhes. Sci. Technol.* **2012**, *26*, 447–461. [CrossRef]
9. Cao, Y.; Wu, Z.; Su, Y.; Xu, Z. Aircraft flight characteristics in icing conditions. *Prog. Aerosp. Sci.* **2015**, *74*, 62–80. [CrossRef]
10. Farzaneh, M. *Atmospheric Icing of Power Networks*; Springer: London, UK, 2008; p. 381.
11. Makkonen, L.; Lehtonen, P.; Hirviniemi, M. Determining ice loads for tower structure design. *Eng. Str.* **2014**, *74*, 229–232. [CrossRef]
12. Kreder, M.; Alvarenga, J.; Kim, P.; Aizenberg, J. Design of anti-icing surfaces: Smooth, textured or slippery. *Nat. Rev. Mater.* **2016**, *1*, 1–15. [CrossRef]
13. Parent, O.; Ilinca, A. Anti-icing and de-icing techniques for wind turbines: Critical review. *Cold Reg. Sci. Technol.* **2011**, *65*, 88–96. [CrossRef]
14. Lamarre, J.-M.; Marcoux, P.; Perrault, M.; Abbott, R.C.; Legoux, J.G. Performance Analysis and Modeling of Thermally Sprayed Resistive Heaters. *J. Therm. Spray Technol.* **2013**, *22*, 947–953. [CrossRef]
15. Lopera-Valle, A.; McDonald, A. Application of Flame-Sprayed Coatings as Heating Elements for Polymer-Based Composite Structures. *J. Therm. Spray Technol.* **2015**, *24*, 1289–1301. [CrossRef]
16. Laforte, C.; Brassard, J.-D.; Volat, C. Extended Evaluation of icephobic coating regarding their field of application. In Proceedings of the International Workshop on Atmospheric Icing of Structures, IWAIS 2019, Reykjavik, Iceland, 23–28 June 2019.
17. Brassard, J.-D.; Laforte, C.; Guerin, F.; Blackburn, C. Icephobicity: Definition and Measurement Regarding Atmospheric Icing. *Adv. Polym. Sci.* **2017**. [CrossRef]
18. Cao, L.; Jones, A.K.; Sikka, V.K.; Wu, J.; Gao, D. Anti-Icing Superhydrophobic Coatings. *Langmuir* **2009**, *25*, 12444–12448. [CrossRef] [PubMed]
19. Wong, T.-S.; Kang, S.H.; Tang, S.K.Y.; Smythe, E.J.; Hatton, B.D.; Grinthal, A.; Aizenberg, J. Bioinspired self-repairing slippery surfaces with pressure-stable omniphobicity. *Nature* **2011**, *477*, 443–447. [CrossRef]

20. Niemelä-Anttonen, H.; Koivuluoto, H.; Tuominen, M.; Teisala, H.; Juuti, P.; Haapanen, J.; Harra, J.; Stenroos, C.; Lahti, J.; Kuusipalo, J.; et al. Icephobicity of Slippery Liquid Infused Porous Surfaces under Multiple Freeze-Thaw and Ice Accretion-Detachment Cycles. *Adv. Mater. Interfaces* **2018**. [CrossRef]
21. Koivuluoto, H.; Stenroos, C.; Kylmälahti, M.; Apostol, M.; Kiilakoski, J.; Vuoristo, P. Anti-icing Behavior of Thermally Sprayed Polymer Coatings. *J. Therm. Spray Technol.* **2017**, *26*, 150–160. [CrossRef]
22. Donadei, V.; Koivuluoto, H.; Sarlin, E.; Vuoristo, P. Icephobic Behaviour and Thermal Stability of Flame-Sprayed Polyethylene Coating: The Effect of Process Parameters. *J. Therm. Spray Technol.* **2020**, *29*, 241–254. [CrossRef]
23. Niemelä-Anttonen, H.; Koivuluoto, H.; Kylmälahti, M.; Laakso, J.; Vuoristo, P. Thermally Sprayed Slippery and Icephobic Surfaces. In Proceedings of the International Thermal Spray Conference, ITSC2018, Orlando, FL, USA, 7–10 May 2018; Azarmi, F., Balani, K., Eden, T., Hussain, T., Lau, Y.-C., Li, H., Shinoda, K., Eds.; pp. 380–384.
24. Lampman, S. Characterization and Failure Analysis of Plastics. In *ASM International Handbooks*; ASM International: Novelty, OH, USA, 2003; p. 482.
25. Lima, C.; de Souza, N.; Camargo, F. Study of Wear and Corrosion Performance of Thermal Sprayed Engineering Polymers. *Surf. Coat. Technol.* **2013**, *220*, 140–143. [CrossRef]
26. Vuoristo, P.; Leivo, E.; Turunen, E.; Leino, M.; Järvelä, P.; Mäntylä, T. Evaluation of Thermally Sprayed and Other Polymeric Coatings for Use in Natural Gas Pipeline Components. In Proceedings of the Thermal Spray, Advancing the Science and Applying the Technology, Orlando, FL, USA, 5–8 May 2003; Moreau, C., Marple, B., Eds.; ASM International: Novelty, OH, USA, 2003; pp. 1693–1702.
27. Gawne, D.T.; Bao, Y.; Zhang, T. Influence of Polymer Composition on The Deposition of UHMWPE Coatings. In Proceedings of the Thermal Spray, Advancing the Science and Applying the Technology, Orlando, FL, USA, 5–8 May 2003; Moreau, C., Marple, B., Eds.; ASM International: Novelty, OH, USA, 2003; pp. 1639–1644.
28. Juuti, P.; Haapanen, J.; Stenroos, C.; Niemelä-Anttonen, H.; Harra, J.; Koivuluoto, H.; Teisala, H.; Lahti, J.; Tuominen, M.; Kuusipalo, J.; et al. Achieving a slippery, liquid-infused porous surface with anti-icing properties by direct deposition of flame synthesized aerosol nanoparticles on a thermally fragile substrate. *Appl. Phys. Lett.* **2017**, *110*, 161603. [CrossRef]
29. Koivuluoto, H.; Stenroos, C.; Ruohomaa, R.; Bolelli, G.; Lusvarghi, L.; Vuoristo, P. Research on icing behavior and ice adhesion testing of icephobic surfaces. In Proceedings of the International Workshop on Atmospheric Icing of Structures, IWAIS 2015, Uppsala, Sweden, 28 June–3 July 2015.
30. Berndt, C.C.; Otterson, D.; Allan, M.L.; Berndt, C.C.; Otterson, D. Polymer Coatings for Corrosion Protection in Biochemical Treatment of Geothermal Residues. *Geotherm. Resour. Counc. Trans.* **1998**, *22*, 425–429.
31. Sugama, T.; Kawase, R.; Berndt, C.C.; Herman, H. An Evaluation of Methacrylic Acid-Modified Poly(Ethylene) Coatings Applied by Flame Spray Technology. *Prog. Org. Coat.* **1995**, *25*, 205–216. [CrossRef]
32. Chen, X.; Gong, Y.; Suo, X.; Huang, J.; Liu, Y.; Li, H. Construction of Mechanically Durable Superhydrophobic Surfaces by Thermal Spray Deposition and Further Surface Modification. *Appl. Surf. Sci.* **2015**, *356*, 639–644. [CrossRef]
33. Bao, Y.; Gawne, D.T.; Zhang, T. Effect of Feedstock Particle Size on the Heat Transfer Rates and Properties of Thermally Sprayed Polymer Coatings. *Trans. Inst. Meter. Finish* **1995**, *73*, 119–124. [CrossRef]
34. Xu, P.; Coyle, T.W.; Pershin, L.; Mostaghimi, J. Superhydrophobic ceramic coating: Fabrication by solution precursor plasma spray and investigation of wetting behavior. *J. Coll. Interface Sci.* **2018**, *523*, 35–44. [CrossRef] [PubMed]
35. Sharifi, N.; Pugh, M.; Moreau, C.; Dolatabadi, A. Developing hydrophobic and superhydrophobic TiO_2 coatings by plasma spraying. *Surf. Coat. Technol.* **2016**, *289*, 29–36. [CrossRef]
36. Niemelä-Anttonen, H.; Kiilakoski, J.; Vuoristo, P.; Koivuluoto, H. Icephobic Performance of Different Surface Designs and Materials. In Proceedings of the International Workshop on Atmospheric Icing of Structures, IWAIS 2019, Reykjavik, Iceland, 23–28 June 2019.
37. Susoff, M.; Siegmann, K.; Pfaffenroth, C.; Hirayama, M. Evaluation of Icephobic Coatings: Screening of Different Coatings and Influence of Roughness. *Appl. Surf. Sci.* **2013**, *282*, 870–879. [CrossRef]
38. Bharathidasan, T.; Kumar, S.; Bobji, M.; Chakradhar, R.; Basu, B. Effect of Wettability and Surface Roughness on Ice-Adhesion Strength of Hydrophilic, Hydrophobic and Superhydrophobic Surfaces. *Appl. Surf. Sci.* **2014**, *314*, 241–250. [CrossRef]

39. Surmeneva, M.; Nikityuk, P.; Hans, M.; Surmenev, R. Deposition of Ultrathin Nano-Hydroxyapatite Films on Laser Micro-Textured Titanium Surfaces to Prepare a Multiscale Surface Topography for Improved Surface Wettability/Energy. *Materials* **2016**, *9*, 862. [CrossRef]
40. He, Y.; Jiang, C.; Cao, X.; Chen, J.; Tian, W.; Yuan, W. Reducing ice adhesion by hierarchical micro-nano-pillars. *Appl. Surf. Sci.* **2014**, *305*, 589–595. [CrossRef]
41. Yeong, Y.H.; Milionis, A.; Loth, E.; Sokhey, J. Self-lubricating icephobic elastomer coating (SLIC) for ultralow ice adhesion with enhanced durability. *Cold Reg. Sci. Technol.* **2018**, *148*, 29–37. [CrossRef]
42. Fu, Q.; Wu, X.; Kumar, D.; Ho, J.W.C.; Kanhere, P.D.; Srikanth, N.; Liu, E.; Wilson, P.; Chen, Z. Development of Sol–Gel Icephobic Coatings: Effect of Surface Roughness and Surface Energy. *ACS Appl. Mater. Interfaces* **2014**, *6*, 20685–20692. [CrossRef] [PubMed]
43. Tejero-Martin, D.; Rezvani Rad, M.; McDonald, A.; Hussain, T. Beyond Traditional Coatings: A Review on Thermal Sprayed Functional and Smart Coatings. *J. Therm. Spray Technol.* **2019**, *28*, 598–644. [CrossRef]
44. Farhadi, S.; Farzaneh, M.; Kulinich, S.A. Anti-icing performance of superhydrophobic surfaces. *Appl. Surf. Sci.* **2011**, *257*, 6264–6269. [CrossRef]
45. Jung, S.; Dorrestijn, M.; Raps, D.; Das, A.; Megaridis, C.M.; Poulikakos, D. Are Superhydrophobic Surfaces Best for Icephobicity? *Langmuir* **2011**, *27*, 3059–3066. [CrossRef] [PubMed]
46. Kulinich, S.; Farzaneh, M. How wetting hysteresis influences ice adhesion strength on superhydrophobic surfaces. *Langmuir* **2009**, *16*, 8854–8856. [CrossRef] [PubMed]
47. Hassan, M.F.; Lee, H.P.; Lim, S.P. The Variation of Ice Adhesion Strength with Substrate Surface Roughness. *Meas. Sci. Technol.* **2010**, *21*, 75701–75709. [CrossRef]
48. Yang, S.; Xia, Q.; Zhu, L.; Xue, J.; Wang, Q.; Chen, Q. Research on the Icephobic Properties of Fluoropolymer-Based Materials. *Appl. Surf. Sci.* **2011**, *257*, 4956–4962. [CrossRef]
49. Zou, M.; Beckford, S.; Wei, R.; Ellis, C.; Hatton, G.; Miller, M.A. Effects of Surface Roughness and Energy on Ice Adhesion Strength. *Appl. Surf. Sci.* **2011**, *257*, 3786–3792. [CrossRef]

© 2020 by the authors. Licensee MDPI, Basel, Switzerland. This article is an open access article distributed under the terms and conditions of the Creative Commons Attribution (CC BY) license (http://creativecommons.org/licenses/by/4.0/).

Article

Characteristics of ZrC Barrier Coating on SiC-Coated Carbon/Carbon Composite Developed by Thermal Spray Process

Bo Ram Kang [1], Ho Seok Kim [1], Phil Yong Oh [1], Jung Min Lee [2], Hyung Ik Lee [2] and Seong Min Hong [1,*]

[1] High-enthalpy Plasma Research Center, Chonbuk National University, 546 Bongdong-ro, Bongdong-eup, Wanju-gun, Jeonbuk 55317, Korea; brkang@jbnu.ac.kr (B.R.K.); hoseok@jbnu.ac.kr (H.S.K.); philoh@jbnu.ac.kr (P.Y.O.)
[2] Agency for Defense Development, Yuseong PO Box 35, Daejeon 305600, Korea; junglee@add.re.kr (J.M.L.); hyungic@add.re.kr (H.I.L.)
* Correspondence: smhong@jbnu.ac.kr; Tel.: +82-10-5427-7740

Received: 7 February 2019; Accepted: 28 February 2019; Published: 5 March 2019

Abstract: A thick ZrC layer was successfully coated on top of a SiC buffer layer on carbon/carbon (C/C) composites by vacuum plasma spray (VPS) technology to improve the ablation resistance of the C/C composites. An optimal ZrC coating condition was determined by controlling the discharge current. The ZrC layers were more than 70 µm thick and were rapidly coated under all spraying conditions. The ablation resistance and the oxidation resistance of the coated layer were evaluated in supersonic flames at a temperature exceeding 2000 °C. The mass and linear ablation rate of the ZrC-coated C/C composites increased by 2.7% and 0.4%, respectively. During flame exposure, no recession was observed in the C/C composite. It was demonstrated that the ZrC coating layer can fully protect the C/C composites from oxidation and ablation.

Keywords: carbon/carbon (C/C) composites; ultra-high temperature ceramic (UHTC); vacuum plasma spray (VPS); ablation resistance

1. Introduction

Carbon/carbon (C/C) composites have low weight, high thermal shock resistance, and good ablation resistance at high temperatures, which facilitate their use in engines, nose tips, leading edges, nozzles, and thermal protection systems for space vehicles operating in severe environments, i.e., during hypersonic flight, atmospheric re-entry, and propulsion [1,2]. However, the working conditions of the abovementioned systems are severe. For example, re-entry vehicles can experience high temperatures of over 2000 °C, which is sustained for several tens to hundreds of seconds with high-pressure air and high-velocity particle erosion. However, C/C composites are vulnerable to oxidation below 400 °C and undergo ablation during high-speed gas jet and particle erosion, which restricts their application in this field. Hence, much effort has been undertaken to improve the ablation resistance of C/C composites for application at higher temperatures and in oxygen-containing atmospheres.

In recent years, extensive research was conducted with the aim of applying ultra-high temperature ceramics (UHTCs) to C/C composites, owing to their high melting point (>3000 °C) and good thermal properties [3–6]. Coating methods are widely used to protect C/C composites owing to the simplicity of the fabrication and integrity with the substrate. UHTCs including transition metal carbides and borides can improve the ablation resistance and ablation properties of C/C composites by doping into the carbon matrix or coating on the composite surface [7–11]. Among UHTCs, ZrC is one of the most promising candidates due to its high melting point of over 3400 °C [12,13]. Additionally, its oxidation product (ZrO_2) has a relatively high melting point of 2700 °C and low vapor pressure; hence, it can form a protective layer on the surface of the C/C composites and reduce the oxygen diffusion rate.

The conventional method for ZrC coating is based on chemical vapor deposition (CVD) [14,15], however, it is complex, expensive, time-consuming, and hazardous to the environment. An alternative low-cost method is chemical vapor infiltration (CVI) [16], however, it has the disadvantage of low efficiency. Both are well developed and widely used to cover the C/C composites by dense ZrC layers, but limit the shape and size of the substrate. Therefore, it is necessary to identify a suitable approach to coat on the C/C composites.

Plasma spraying is one of the most promising coating methods because it is economical and can be easily applied on an industrial scale [17–19]. It is also a suitable technique for depositing carbides with high melting points such as ZrC, HfC, etc. [20,21]. In the VPS (vacuum plasma spray) process or in other plasma spray processes in the atmosphere, ZrC powder is oxidized to result in the formation of zirconium oxides [22–25]. Therefore, VPS coatings are especially preferred because of their high purity with low porosity and high deposition rate without the formation of oxides [26]. In VPS, the coating layer is formed by melting the constituent particles at a very high temperature and by colliding them with the base material. Therefore, the inherent properties of both components, i.e., the molten particles and base substrate, are preserved. However, as numerous parameters contribute to the formation of a high-quality coating layer, it is important to determine the optimal process parameters. Among the operational parameters, it is well known that the velocity and temperature of the plasma jet, which can be controlled mainly by the discharge current, are closely associated with the phase change, thickness, and porosity. As mentioned above, although the coating process using VPS has many advantages, few research articles have reported on UHTCs coatings prepared using the VPS technique.

Therefore, in this work, we investigated the discharge current effects for a ZrC coating layer on a C/C composite and evaluated the thermal ablation performance of the coating. Scanning electron microscopy (SEM) and X-ray diffraction (XRD) analyses were performed to analyze the coating layer properties. Furthermore, an ablation test was conducted to determine the ablation resistance of coating layers fabricated under various current conditions of VPS.

2. Materials and Methods

2.1. ZrC Coating Process on C/C Composites

Disc-shaped C/C composites (Dai-Yang Industry Co.) with a 5 mm thickness and a 30 mm diameter were used as the substrates. To reduce the difference in the thermal expansion coefficient between ZrC and the carbon composites, a thick SiC layer (approximately 30 µm) was fabricated on the top surface of the C/C composite by chemical vapor reaction (CVR) [27]. ZrC powders (D50 = 8 µm; purity >99.9%, Avention, Korea) were used to coat the ZrC layer using a VPS system (Oerlikon Metco AG, Wohlen, Switzerland) with a F4-VB gun (Oerlikon Metco AG, Wohlen, Switzerland). The ZrC raw powders exhibited irregular shapes (Figure 1).

Prior to the coating process, the substrates were installed in the chamber and aligned perpendicular to the gun. Figure 2 shows the three main steps involved in the coating process. The first step is pre-heating the substrates, which increases the liquidity of the substrate surface and improves the adhesion flexibility of the surface with the molten ZrC droplets in the next coating step. Moreover, impurities on the surface are removed in this step. The splat shape of the powder varies depending on the pre-heating temperature. As the temperature increases, the radius of the splats increases, however the droplets spread too wide, thereby hindering the formation of the coating. Therefore, it is important to choose the appropriate pre-heating temperature. The distance between the substrate and plasma gun was fixed at 210 mm and the pre-heating was repeated along an S-shaped path. The movement of the plasma gun from the top left to the bottom right of the substrate and then back to the top left was defined as one cycle, with a total of five cycles performed. Throughout the pre-heating process, the surface temperature of the substrate was maintained below 900 °C.

Figure 1. Field emission scanning electron microscopy (FE-SEM) morphology of ZrC powder.

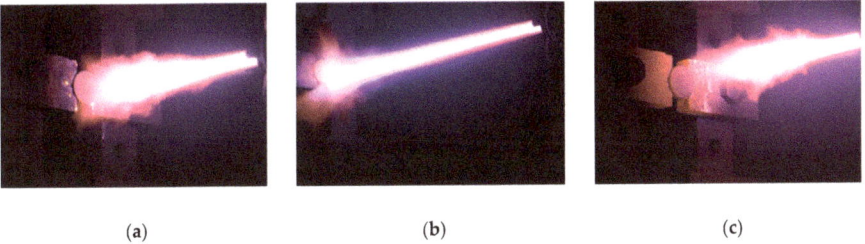

Figure 2. The coating process: (**a**) Pre-heating, (**b**) coating, and (**c**) post-heating.

In the second step, the coating process (Figure 2b), the ZrC powder from the injector was melted by the plasma flame, and the droplets were attached on the substrates with a high speed. In this step, the distance between the substrates and plasma gun was adjusted to 350 mm. Ar and H_2 were used as the plasma forming gases, and Ar was also used as the powder carrier gas. The powder feeding rate was 4.5 g/min. To obtain a coating layer with uniform thickness, coating was performed over an S-shaped path, similar to the pre-heating process. Coating was performed at 10 mm intervals and in all, 20 coatings were applied. The current was varied as 500 A, 550 A, 600 A, 650 A, and 700 A in the experiment, and the coated sample was denoted as ZS-1, ZS-2, ZS-3, ZS-4, and ZS-5, respectively. Table 1 shows the detailed process conditions.

Finally, the specimens' temperatures sharply decreased after the coating process. At this moment, post-heating was applied to the specimens to prevent crack/defect formation, and to prevent exfoliation of the coating layer due to the difference in the thermal expansion coefficient between the substrates and the ZrC coating. Post-heating treatment was performed under a lower current (500 A) compared to the current used in the pre-heating and coating steps, so as to prevent damage to the coated specimens. The total time taken for coating was around 5 min.

Table 1. Details of the vacuum plasma spraying (VPS) parameters for ZrC coating.

	Parameters	
Pre/post-heating	Spraying current, A	500
	Ar gas flow, NLPM [1]	30
	H_2 gas flow, NLPM	2
	Chamber Pressure, mbar	100
	Spraying distance, mm	210

Table 1. Cont.

	Parameters	
Coating	Spraying current, A	500–700
	Ar gas flow, NLPM	50
	H_2 gas flow, NLPM	10
	Feeding rate, g/min	4.5
	Chamber Pressure, mbar	50
	Spraying distance, mm	350

[1] Normal liter per minute.

2.2. Coating Characterization

After cutting the five coated samples, their surfaces were polished using 1 μm of diamond paste to analyze the characteristics of the coating layer. Phase analysis of the coating was conducted by XRD (D8 Advance, Bruker, USA) at 40 kV, with Cu-Kα radiation, and a scanning speed of 4 °/min. The cross-sectional microstructure of the coating was determined by field emission scanning electron microscopy (FE-SEM; SU-8030, Hitachi, Japan), and the components of the coating layer were analyzed by energy dispersive X-ray spectroscopy (EDS; X-MaxN80, Horiba, Japan). Table 2 shows the results of the analysis for the coating layer. The thickness and porosity were determined using an image analyzer by referring to the FE-SEM images and by porosity measurements based on ASTM E-2109 (Test Methods for Determining Area Percentage Porosity in Thermal Sprayed Coatings). Additionally, the coating layer's bonding strength and porosity were analyzed. Finally, the ablation test was carried out for the ZrC-coated sample, which was selected based on the best result in metallurgical analysis. Before the ablation test, an adhesion test of the selected sample was also carried out by the universal testing machine (UTM; 5982, Instron, USA). Figure 3 shows a schematic of the adhesion test with the sample fixed with glue (Fusionbond 370, Hernon, USA) between the bars.

Table 2. Ablation properties of the carbon/carbon (C/C) composite with or without ZrC coating.

Samples	Mass Ablation Rate ($\times 10^{-2}$ g/s)	Linear Ablation Rate ($\times 10^{-3}$ mm/s)
C/C	3.21	7.73
ZrC/SiC-coated C/C	−2.67	−3.76

Figure 3. A schematic of the adhesion test.

2.3. Ablation Test with Supersonic Flame

The ZrC coating layer was expected to protect the C/C composites in a high-temperature oxidizing environment. Therefore, the weights and thicknesses of the sample before and after ablation were measured by an ablation test, and the ablation rate was calculated. The ablation characteristics of the coated and uncoated samples were compared, and the performance of the coating layer was verified based on the results.

Among the five coated samples (ZS-1~ZS-5), the sample with the best characteristics was selected for the ablation test. A high-velocity oxygen fuel (HVOF) system equipped with a diamond jet (DJ) gun (DJH2700 gun; Oerlikon Metco, United States of America) was used. The DJ gun uses a combination of oxygen, fuel gas, and air to generate a high-pressure flame with uniform temperature distribution. A pre-mixed fuel gas (typically, propane, methane, propylene, or oxygen) and oxygen come into contact with the supplied air to generate a high-temperature combustion gas. The temperature of the generated flame approaches 2730 °C, and the flame is accelerated through convergent/divergent nozzles to create a supersonic flame. In this work, commercial liquefied petroleum gas (LPG) was used as the fuel. The pressure and flow velocity were 10.3 bar and 22 L/min for oxygen, 6.2 bar and 7.5 L/min for the fuel, and 7.2 bar and 58 L/min for air. As shown in Figure 4, the specimen was installed 60 mm away from the nozzle and was aligned perpendicular to the nozzle. The ablation test was conducted for 30 s at temperatures higher than 2230 °C in open air. During ablation, the surface temperature of the specimen was measured using a two-color pyrometer. We used a digimatic micrometer (Mitutoyo), a non-contact 3D surface measuring system (IFM G4, Alicona), and a 0.1 mg-precise electronic balance for before and after test evaluation of the recession and mass loss of the test samples. The linear and mass ablation rates of the sample were calculated using the following equations:

$$R_m = \frac{\Delta m}{t} \tag{1}$$

$$R_l = \frac{\Delta l}{t} \tag{2}$$

where, R_m is the mass ablation rate, Δm is the mass change before and after ablation, R_l is the linear ablation rate, Δl is the change of the thickness at ablation region, and t is the ablation time.

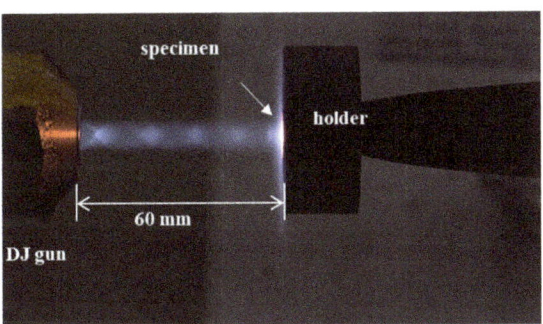

Figure 4. Test specimen mounted on holder.

3. Results and Discussion

3.1. Microstructure and Phase of the As-Sprayed Coatings

Figure 5 shows the samples' FE-SEM cross-sections before and after the ZrC coating layers by different conditions. Figure 5a shows that SiC is well bonded on the C/C composite substrate, with a 30 μm thickness. The coated samples (Figure 5b–f) show different coating layer characteristics depending on the discharge currents. ZrC is uniformly coated on top of SiC without any loss of SiC, however some cracks formed by in the coating layer of samples during the cutting. The thickness of the coating layers gradually increases from 75 to 122 μm as the discharge current increases. While the thickness increases, more cracks were observed. As cracks formed during the cutting for cross-sectional observation due to the accumulated thermal stress, a partial delamination from the substrate was observed in a thicker case, as shown in Figure 5f.

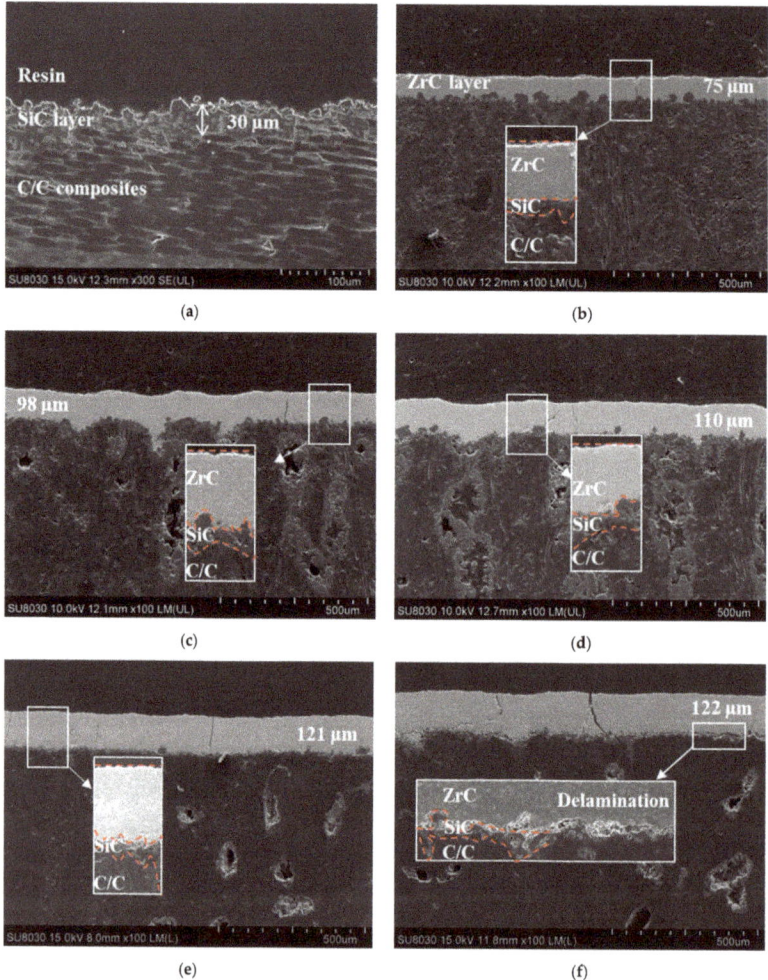

Figure 5. Cross-section morphology of the substrate and as-sprayed coating: (**a**) Substrate, (**b**) ZS-1, (**c**) ZS-2, (**d**) ZS-3, (**e**) ZS-4, and (**f**) ZS-5.

Figure 6 shows high-magnification FE-SEM images, detailing the interface between the ZrC and SiC layers. The light gray region represents the ZrC layer, while the dark gray region represents the SiC layer. There is no mixed region between the ZrC layer and the C/C substrate, hence, each layer is distinct. Figure 7 shows the results of the porosity measurement. Although there is no significant change in the porosity, the porosity slightly increases with the current with the exception of ZS-2. The discharge current could significantly affect the velocities and temperatures of the particles as well as the thermal/physical characteristics of the plasma flow. If the current increases, the discharge power, which can bring about a higher gas temperature of the plasma jet, also increases, therefore, more powders can be melted. Splashing occurs due to poor spreading of the particles on the substrate surface. Therefore, the porosity increases as the molten particles splash in the substrate under a high current. For the ablation test, the above results were considered and the ZS-2 sample was selected as the one with the best characteristics. Figure 8 shows the results of the phase analysis of the coated sample. Figure 8a shows the EDS mapping results of the ZS-2 sample. In the upper part of the figure, light blue represents Zr, green represents Si, and red represents C. Figure 8b shows the XRD diffraction pattern of the coated sample fabricated using VPS. As shown in Figure 8b, it is clear that there are strong peaks corresponding to ZrC, while diffraction peaks of carbon are not detected, indicating that the thick ZrC layer is formed on the C/C composite. Meanwhile, narrow and sharp peaks indicate good crystallization of the ZrC phase. The peaks at $2\theta = 33°$, $36°$, and $56°$ match well with those of ZrC corresponding to lattice planes of (111), (200), and (220) according to the Joint Committee on Powder Diffraction Standards (JCPDS) database [11-0110]. From the result of the XRD and EDS analyses, SiC and ZrC are well identified, and confirm that ZrC is well coated on the substrate without the formation of oxides or the influx of impurities.

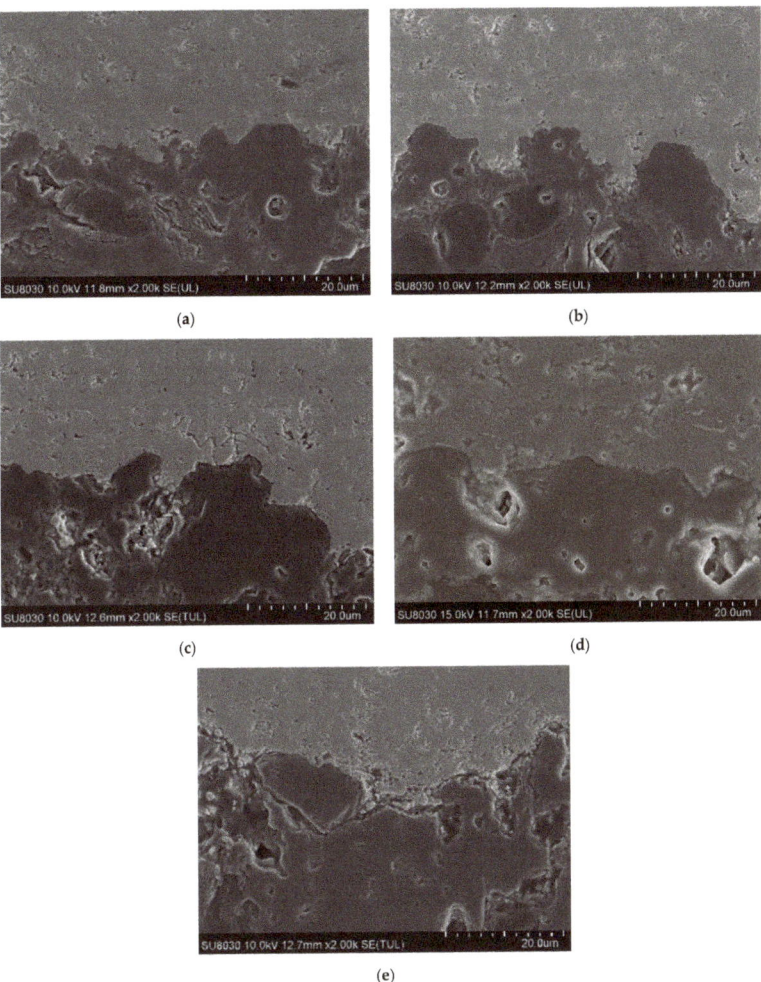

Figure 6. FE-SEM morphologies of the ZrC/SiC interfaces: (**a**) ZS-1, (**b**) ZS-2, (**c**) ZS-3, (**d**) ZS-4, and (**e**) ZS-5.

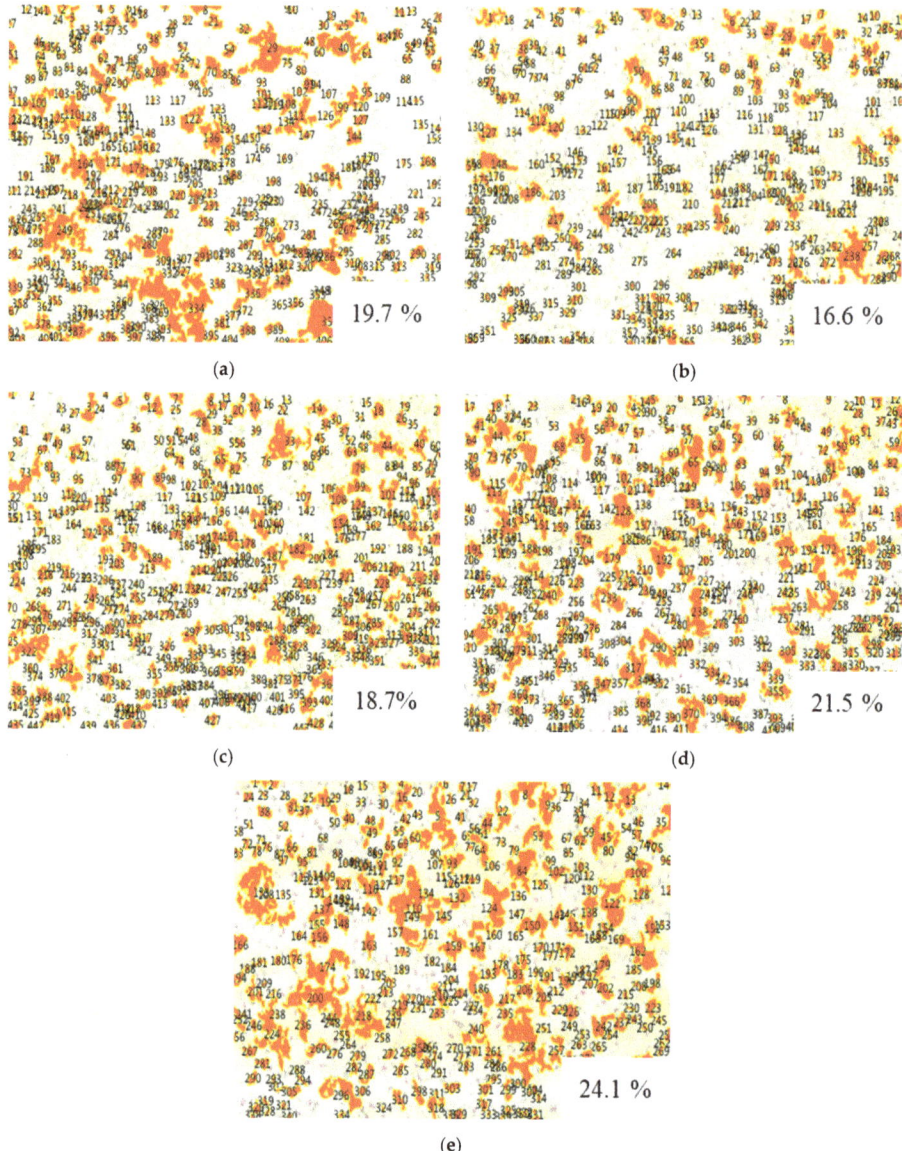

Figure 7. Porosity of coating layer: (a) ZS-1, (b) ZS-2, (c) ZS-3, (d) ZS-4, and (e) ZS-5.

To perform the ablation test, we selected the ZS-2 condition because of its obtained low porosity and uniform thickness without cracks and delamination. Using this condition, we increased the number of coating cycles to obtain thicker layers of up to 163 µm, which should be useful for prolonged ablation. Figure 9 shows the results of the adhesion test on the ZS-2 sample. Figure 9a shows tensile strength and Figure 9b shows the tested sample ZS-2. As a result, the ZrC layer was not separated with the SiC layer, however it was separated with the P1. Furthermore, the ZrC layer was not peeled from the substrate. Based on these results, it can be seen that ZrC is well bonded with SiC.

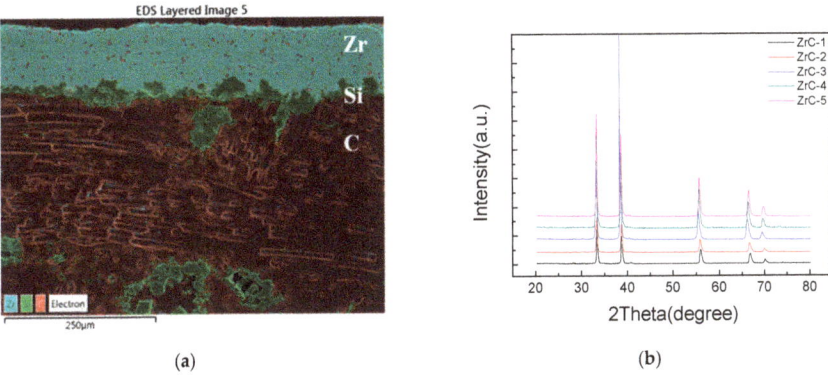

Figure 8. Phase analysis result: (**a**) Energy dispersive X-ray spectroscopy (EDS) mapping analysis data of ZS-2 and (**b**) XRD analysis data of as-sprayed coating.

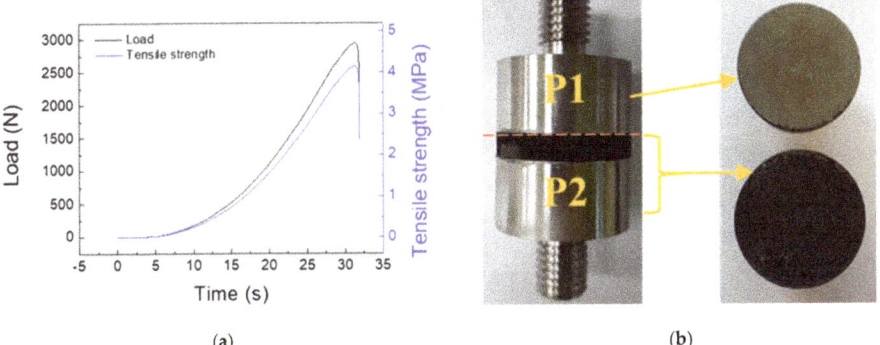

Figure 9. Adhesion test result of ZS-2: (**a**) Tensile strength and (**b**) morphologies after the adhesion test on the ZS-2 sample.

3.2. Ablation Properties

Table 2 shows the ablation properties of the uncoated and ZrC-coated C/C composites. Note that there is an obvious weight loss in the ablation process for the C/C composite, while the coated sample gains weight. For the specimen without the coating, the weight decreases by 0.95 g. The thickness of the specimen also decreases however, it was difficult to accurately measure the ablation rate as the diameter of the flame was smaller than that of the specimen and consequently, ablation was not uniform. Therefore, the ablation rate of the specimen was calculated using a non-contact 3D surface measuring system (IFM G4, Alicona). Figure 10a shows the surface image of the ablated specimen, from which the ablated length was calculated using the difference in heights before and after the ablation test, which is shown in Figure 10b. The most ablated part was measured and determined as the linear ablation rate. On the other hand, for the coated specimen, the weight and thickness increased by 0.8 g and 113 µm, respectively. Figure 11 shows the results of XRD analysis of the specimen surface after the ablation test and ZrO_2 component. Only the oxidation of ZrC is observed. It appears that the outer ZrC layer can be transformed into ZrO_2, which results in weight and volume increases due to the formation of the oxide. The ablation test was performed for 30 s, and the maximum surface temperatures of the coated and uncoated specimens were measured to be 2052 °C and 2275 °C, respectively.

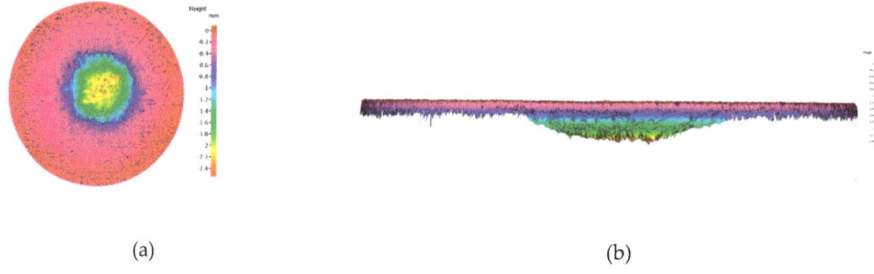

(a)   (b)

Figure 10. Ablation surface images for (**a**) top and (**b**) side.

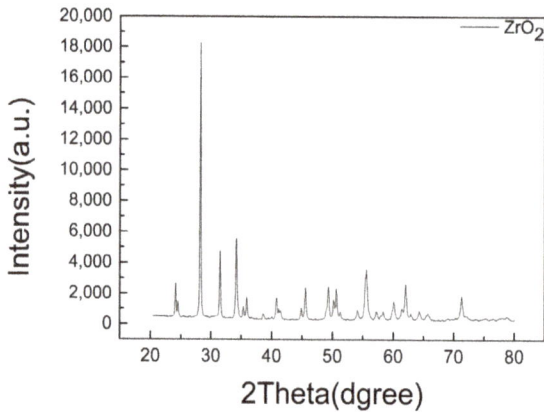

Figure 11. XRD analysis of ZrC coating for C/C composite after ablation for 30 s.

3.3. Morphology Analysis of the Coatings After Ablation

Figure 12 shows images of the uncoated and coated C/C composites before and after ablation. In the uncoated C/C composite, the center is subjected to the maximum ablation. Furthermore, the surface color of the coated specimen changes from black to white by oxidation.

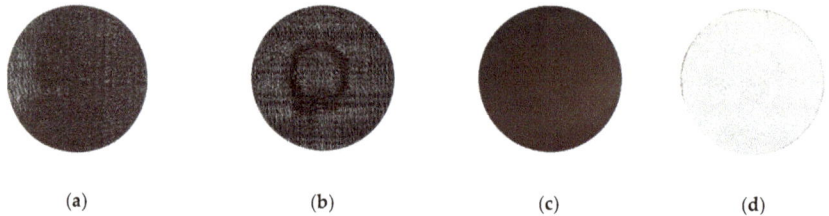

(a)   (b)   (c)   (d)

Figure 12. Morphologies before (**a**), (**c**) and after (**b**), (**d**) the ablation test; (**a**), (**b**) uncoated and (**c**), (**d**) coated specimens.

Figure 13a,c show the surface morphology of the uncoated C/C composite before ablation, while Figure 13b,d show the surface morphology after ablation. After ablation, several large pores generated on the surface, which show irregular patterns as seen in Figure 13b. Additionally, the spaces between the fibers are filled with carbon matrices (Figure 13c), however after ablation, the carbon matrices disappear and the carbon fibers are exposed with a decreased thickness (Figure 13d). Furthermore, the fibers show a tapered shape that indicates that the carbon matrices are first oxidized, followed by the fibers, due to which the tips of the fibers are deformed into a needle shape (Figure 13d).

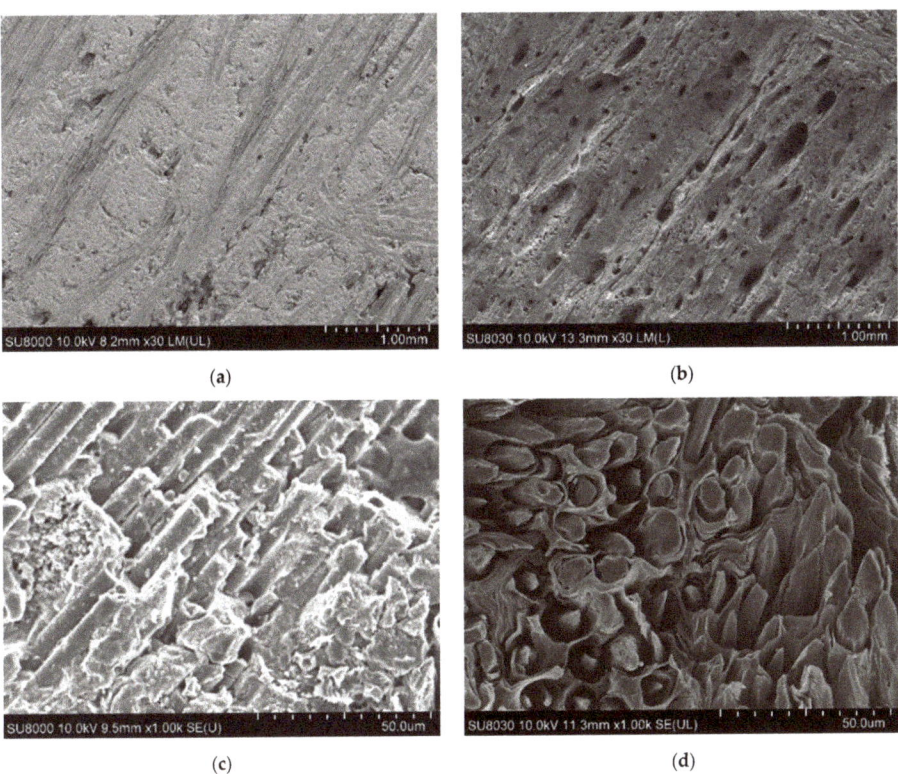

Figure 13. Before (**a**) and(**c**), after (**b**) and (**d**), ablation surface morphology of the C/C composite.

Figure 14 shows images of the specimen's surface before and after the ablation test. Figure 14a shows the top of the ZrC coating layer before ablation, wherein cracks and pores are not observed. Figure 14b shows the center area in the ablation specimen, from which it can be seen that the ZrO$_2$ layer has completely melted. This is confirmed in Figure 12d, wherein the cracks seem fine on the entire surface, from the center of the specimen after ablation, and microcracks and micropores are observed from the surface, as shown in Figure 14b. It appears that such cracks and defects occurred due to the difference in the thermal expansion coefficient caused by a sudden thermal shock. Furthermore, it seems that many microcracks and micropores generated on the specimen surface because the ZrO$_2$ formed on the specimen surface underwent a phase change into tetragonal zirconia at a high temperature, which further changes into monoclinic zirconia upon cooling after completion of the ablation test. Therefore, as mentioned, cracks seem to have formed due to the volume change and rapid cooling caused by the two phase changes, and pores seem to have formed due to CO or CO$_2$ gas, which was generated when ZrC reacted with oxygen.

Figure 15 shows FE-SEM images and EDS mapping analysis of the specimen's cross-section after the ablation test. Figure 15a shows the cross-section of the coated specimen after the ablation test. Part of the coating layer appears to be exfoliated when the specimen was cut for observation. However, the bottom of Figure 15a shows that the remaining ZrC layer is still well bonded on the SiC layer. At the top of the coating layer, a uniform oxide layer with over 30 μm thickness formed. From XRD analysis (Figure 11), the coating layer is covered by ZrO$_2$. Additionally, the ZrC layer and substrate are well bonded at the interface without exfoliation after the ablation test. Figure 15b,c show the image and results of the EDS line scan profile and point analysis, respectively. The upper and middle parts of the coating layer, and the substrate beneath the coating layer were selected for analysis. The white phase in spectrum 1 is primarily ZrO$_2$ produced by ZrC oxidation, and the light gray phase in spectrum 2 and spectrum 3 is mainly ZrC and ZrO$_2$. The gray phase in spectrum 4 is mainly SiC with no SiO$_2$ detected. The proportion of oxygen decreases toward the substrate. SiO$_2$ is not observed as the oxygen could not spread to the SiC layer that forms an interface with ZrC. Figure 15c shows the results of the sample's line scan analysis. Carbon peaks are observed, because the resin flows through the gap in the middle of the coating layer. Oxygen peaks are observed with a high intensity at 70–80 μm depth, which weakens thereafter. Based on the EDS and previous XRD results, it is clear that the ZrO$_2$ protective film on the surface of the coating layer effectively prevents oxygen diffusion from ZrC to the C/C composite. Furthermore, the ZrO$_2$ layer serves as a barrier to retard thermal diffusion and thus, reduced heat transfer to the underlying C/C composites.

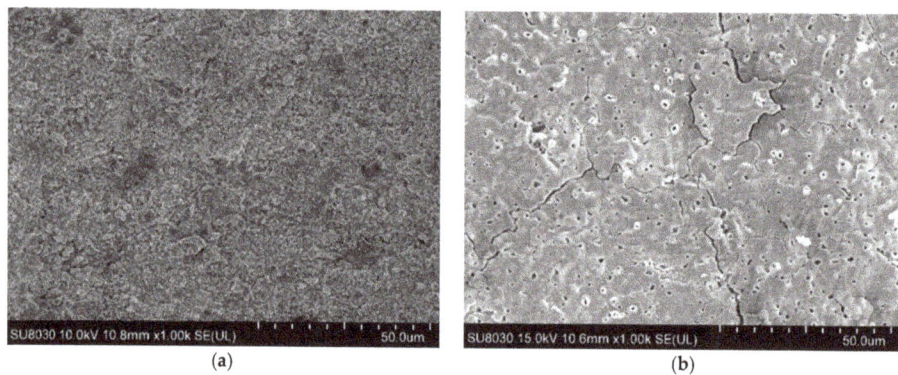

Figure 14. (a) Surface morphologies of the ZrC-coated sample before, (b) and after the ablation test.

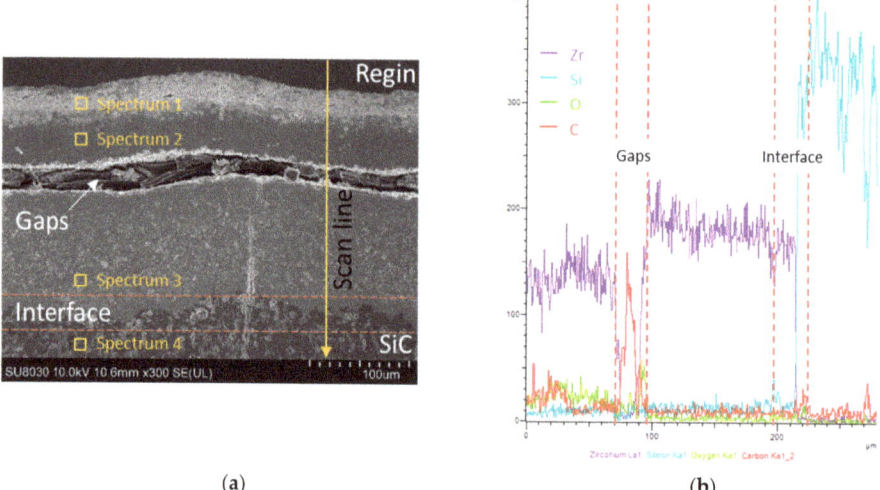

Figure 15. *Cont.*

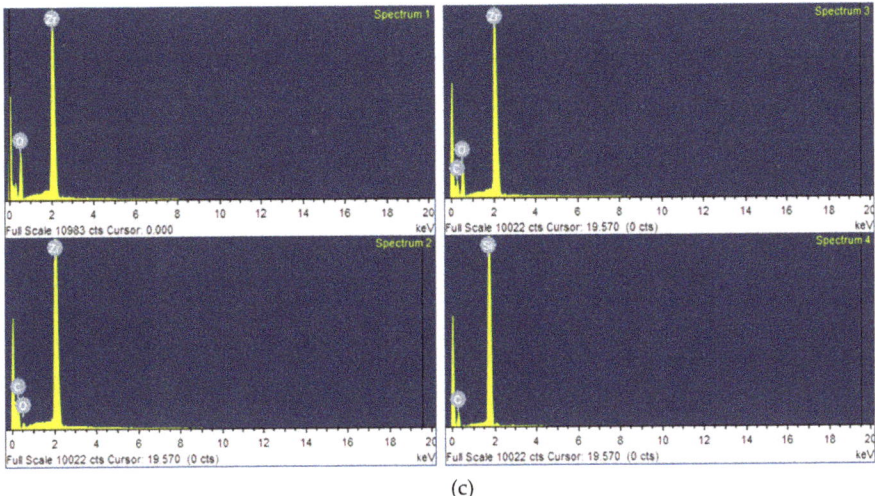

(c)

Figure 15. Cross-sectional FE-SEM images and EDS analysis of the ZrC-coated samples: (**a**) Cross-section image, (**b**) Line scan profile, and (**c**) Point analysis.

4. Conclusions

We developed an optimal ZrC coating process by changing the discharge current in a VPS system for improving thermal stability of a C/C composite. The coating was performed at different current values that ranged from 500 to 700 A, and the ablation test was conducted for the best condition of the ZrC coating using a HVOF system. Thick ZrC layers were coated on the C/C composite, where SiC layers formed by CVR served as the interlayer between them. By increasing the discharge current, the coated thickness and porosity also increased. At a higher current condition, cracks and delamination of the ZrC layer commenced. During the ablation test, the C/C composite was completely protected from thermal oxidation by the thick ZrC layer which formed ZrO_2. After the ablation test, although cracks and pores were observed from the surface of the coating layer, no delamination was observed at the coating layer interface. Phase analysis by XRD and EDS indicated that the ZrO_2 formed at the top of the coating layer was transferred into the C/C composites. Thus, the coating process using VPS effectively protected a C/C composite in an ablation environment; moreover, a detached coating was not detected between the ZrC coating and substrate, indicating good adherence between them.

Author Contributions: Conceptualization, B.R.K., H.S.K., J.M.L., H.I.L., and S.M.H.; methodology, B.R.K. and H.S.K.; validation, B.R.K., H.S.K., and P.Y.O.; formal analysis, B.R.K. and H.S.K.; investigation, B.R.K., H.S.K., J.M.L., and H.I.L.; resources, B.R.K., H.S.K., and P.Y.O.; data curation, B.R.K. and H.S.K.; writing—original draft preparation, B.R.K.; writing—review and editing, S.M.H.; visualization, B.R.K.; supervision, S.M.H.; project administration, B.R.K., H.S.K., and S.M.H.; funding acquisition, J.M.L. and H.I.L.

Funding: This work was supported by the Korean Government (Defense Acquisition Program Administration, DAPA) through Agency for Defense Development (contract number is UD160084BD).

Acknowledgments: The authors would like to thank Prof. S.Y Moon for sharing his experimental ideas for coating and allowing us to reference them in this work.

Conflicts of Interest: The authors declare no conflict of interest.

References

1. Fitzer, E. The future of carbon-carbon composites. *Carbon* **1987**, *25*, 163–190. [CrossRef]
2. Zaman, W.; Li, K.Z.; Ikram, S.; Li, W.; Zhang, D.S.; Guo, L.J. Morphology, thermal response and anti-ablation performance of 3D-four directional pitch-based carbon/carbon composites. *Corros. Sci.* **2012**, *61*, 134–142. [CrossRef]
3. Deng, J.; Cheng, L.; Hong, Z.; Su, K.; Zhang, L. Thermodynamics of the production of condensed phases in the chemical vapor deposition process of zirconium diboride with ZrCl4–BCl3–H2 precursors. *Thin Solid Films* **2012**, *520*, 2331–2335. [CrossRef]
4. Justin, J.; Jankowiak, A. Ultra high temperature ceramics: Densification, properties and thermal stability. *AerospaceLab* **2011**, *3*, 1–11.
5. Levine, S.R.; Opila, E.J.; Halbig, M.C.; Kiser, J.D.; Singh, M.; Salem, J.A. Evaluation of ultra-high temperature ceramics for aeropropulsion use. *J. Eur. Ceram. Soc.* **2002**, *22*, 2757–2767. [CrossRef]
6. Wuchina, E.; Opeka, M.; Causey, S.; Buesking, K.; Spain, J.; Cull, A.; Routbort, J.; Guitierrez-Mora, F. Designing for ultrahigh-temperature applications: the mechanical and thermal properties of HfB$_2$, HfCx, HfNx and αHf (N). *J. Mater. Sci.* **2004**, *39*, 5939–5949. [CrossRef]
7. Krishnarao, R.; Alam, M.Z.; Das, D.J.C.S. In-situ formation of SiC, ZrB2-SiC and ZrB2-SiC-B4C-YAG coatings for high temperature oxidation protection of C/C composites. *Corros. Sci.* **2018**, *141*, 72–80. [CrossRef]
8. Wang, J.S.; Li, Z.P.; Ao, M.; Xu, Z.H.; Liu, L.; Hu, J.; Peng, W.Z. Effect of doped refractory metal carbides on the ablation mechanism of carbon/carbon composites. *New Carbon Mater.* **2006**, *21*, 9–13.
9. Wei, H.; Min, L.; Chunming, D.; Xuezhang, L.; Dechang, Z.; Kesong, Z. Ablation resistance of APS sprayed mullite/ZrB2-MoSi2 coating for carbon/carbon composites. *Rare Metal Mater. Eng.* **2018**, *47*, 1043–1048. [CrossRef]
10. Xuetao, S.; Kezhi, L.; Hejun, L.; Hongying, D.; Weifeng, C.; Fengtao, L. Microstructure and ablation properties of zirconium carbide doped carbon/carbon composites. *Carbon* **2010**, *48*, 344–351. [CrossRef]
11. Yang, Y.; Li, K.; Zhao, Z.; Liu, G. HfC-ZrC-SiC multiphase protective coating for SiC-coated C/C composites prepared by supersonic atmospheric plasma spraying. *Ceram. Int.* **2017**, *43*, 1495–1503. [CrossRef]
12. Grossman, L.N. High-temperature thermophysical properties of zirconium carbide. *J. Am. Ceram. Soc.* **1965**, *48*, 236–242. [CrossRef]
13. Sara, R.V. The system zirconium—carbon. *J. Am. Ceram. Soc.* **1965**, *48*, 243–247. [CrossRef]
14. Wang, S.L.; Li, K.Z.; Li, H.J.; Zhang, Y.L.; Feng, T. Structure evolution and ablation behavior of ZrC coating on C/C composites under single and cyclic oxyacetylene torch environment. *Ceram. Int.* **2014**, *40*, 16003–16014. [CrossRef]
15. Wang, S.L.; Li, K.Z.; Li, H.J.; Zhang, Y.L.; Wang, Y.J. Effects of microstructures on the ablation behaviors of ZrC deposited by CVD. *Surf. Coat. Technol.* **2014**, *240*, 450–455. [CrossRef]
16. Zhou, H.; Ni, D.; He, P.; Yang, J.; Hu, J.; Dong, S. Ablation behavior of C/C-ZrC and C/SiC-ZrC composites fabricated by a joint process of slurry impregnation and chemical vapor infiltration. *Ceram. Int.* **2018**, *44*, 4777–4782. [CrossRef]
17. Kobayashi, A.; Yano, S.; Kimura, H.; Inoue, A. Fe-based metallic glass coatings produced by smart plasma spraying process. *Mater. Sci. Eng., B* **2008**, *148*, 110–113. [CrossRef]
18. Yugeswaran, S.; Selvarajan, V.; Vijay, M.; Ananthapadmanabhan, P.V.; Sreekumar, K.P. Influence of critical plasma spraying parameter (CPSP) on plasma sprayed alumina–titania composite coatings. *Ceram. Int.* **2010**, *36*, 141–149. [CrossRef]
19. Lemlikchi, S.; Martinsson, J.; Hamrit, A.; Djelouah, H.; Asmani, M.; Carlson, J. Ultrasonic Characterization of Thermally Sprayed Coatings. *J. Therm. Spray Technol.* **2019**, *28*, 391–404. [CrossRef]
20. Wu, H.; Li, H.J.; Fu, Q.G.; Yao, D.J.; Wang, Y.J.; Ma, C.; Han, Z.H. Microstructures and ablation resistance of ZrC coating for SiC-coated carbon/carbon composites prepared by supersonic plasma spraying. *J. Therm. Spray Technol.* **2011**, *20*, 1286–1291. [CrossRef]
21. Yang, Y.; Li, K.; Li, H. Ablation Behavior of HfC Coating Prepared by Supersonic Plasma Spraying for SiC-Coated C/C Composites. *Adv. Compos. Lett.* **2015**, *24*, 113–118. [CrossRef]
22. Wen, B.; Ma, Z.; Liu, Y.; Wang, F.; Cai, H.; Gao, L. Supersonic flame ablation resistance of W/ZrC coating deposited on C/SiC composites by atmosphere plasma spraying. *Ceram. Int.* **2014**, *40*, 11825–11830. [CrossRef]

23. Jia, Y.; Li, H.; Feng, L.; Sun, J.; Li, K.; Fu, Q. Ablation behavior of rare earth La-modified ZrC coating for SiC-coated carbon/carbon composites under an oxyacetylene torch. *Corros. Sci.* **2016**, *104*, 61–70. [CrossRef]
24. Hu, C.; Ge, X.; Niu, Y.; Li, H.; Huang, L.; Zheng, X.; Sun, J. Influence of oxidation behavior of feedstock on microstructure and ablation resistance of plasma-sprayed zirconium carbide coating. *J. Therm. Spray Technol.* **2015**, *24*, 1302–1311. [CrossRef]
25. Jia, Y.; Li, H.; Fu, Q.; Zhao, Z.; Sun, J. Ablation resistance of supersonic-atmosphere-plasma-spraying ZrC coating doped with ZrO2 for SiC-coated carbon/carbon composites. *Corros. Sci.* **2007**, *123*, 40–54. [CrossRef]
26. Niu, Y.R.; Liu, X.Y.; Ding, C.X. Comparison of silicon coatings deposited by vacuum plasma spraying (VPS) & atmospheric plasma spraying (APS). *Mater. Sci. Forum.* **2006**, *510*, 802–805.
27. Kowbel, W.; Withers, J.; Ransone, P. CVD and CVR silicon-based functionally gradient coatings on C C composites. *Carbon* **1995**, *33*, 415–426. [CrossRef]

© 2019 by the authors. Licensee MDPI, Basel, Switzerland. This article is an open access article distributed under the terms and conditions of the Creative Commons Attribution (CC BY) license (http://creativecommons.org/licenses/by/4.0/).

Article

Effect of Adjusted Gas Nitriding Parameters on Microstructure and Wear Resistance of HVOF-Sprayed AISI 316L Coatings

Pia Kutschmann [1,*], Thomas Lindner [1], Kristian Börner [2], Ulrich Reese [2] and Thomas Lampke [1]

1. Materials and Surface Engineering Group, Institute of Materials Science and Engineering, Chemnitz University of Technology, D-09107 Chemnitz, Germany; th.lindner@mb.tu-chemnitz.de (T.Li.); thomas.lampke@mb.tu-chemnitz.de (T.La.)
2. Härterei Reese Chemnitz GmbH & Co. KG, 09117 Chemnitz, Germany; KBoerner@haerterei.com (K.B.); UReese@haerterei.com (U.R.)
* Correspondence: pia.kutschmann@mb.tu-chemnitz.de; Tel.: +49-371-531-30052

Received: 13 May 2019; Accepted: 29 May 2019; Published: 30 May 2019

Abstract: Gas nitriding is known as a convenient process to improve the wear resistance of steel components. A precipitation-free hardening by low-temperature processes is established to retain the good corrosion resistance of stainless steel. In cases of thermal spray coatings, the interstitial solvation is achieved without an additional surface activation step. The open porosity permits the penetration of the donator media and leads to a structural diffusion. An inhomogeneous diffusion enrichment occurs at the single spray particle edges within the coating's microstructure. A decreasing diffusion depth is found with increasing surface distance. The present study investigates an adjusted process management for low-temperature gas nitriding of high velocity oxy-fuel-sprayed AISI 316L coatings. To maintain a homogeneous diffusion depth within the coating, a pressure modulation during the process is studied. Additionally, the use of cracked gas as donator is examined. The process management is designed without an additional surface activation step. Regardless of surface distance, microstructural investigations reveal a homogeneous diffusion depth by a reduced processing time. The constant hardening depth allows a reliable prediction of the coatings' properties. An enhanced hardness and improved wear resistance is found in comparison with the as-sprayed coating condition.

Keywords: thermal spraying; high velocity oxy-fuel (HVOF); S-phase; expanded austenite; 316L; stainless steel; thermochemical treatment; hardening; gas nitriding

1. Introduction

Thermochemical treatment is a common process for functionalizing the surface of stainless steels to improve their wear characteristics. Depending on the temperature-time regime during the process, the nitrogen enrichment leads to the formation of precipitates at temperatures above 500 °C, or an interstitial solid solution below 450 °C. Supersaturation of the austenitic matrix decisively expands the lattice parameters. This phase condition is named S-phase or expanded austenite [1–3].

Hardening of austenitic stainless steels comprises an initial surface activation step in order to remove the passive layer. Special equipment is required to ensure absence of repassivation during the diffusion enrichment. This significantly increases the complexity of the process management and often causes high costs. In contrast, classical thermochemical treatments possess clear economic benefits; in particular, the high flexibility and the relatively low processing costs are responsible for larger-scale industrial application of the gas nitriding process [4,5].

Different parameter settings related to bulk materials aim to increase process efficiency in order to reduce processing time or enhance hardness depth. Aleekseeva et al. applied a high-pressure gas nitriding process at a high temperature of 1150 °C, a high pressure of 150 MPa and a duration of 3 h. The diffusion layer in martensitic steels reaches 1.5–2 mm thickness [5]. In contrast, Wolowiec-Korecka et al. evaluated the low-pressure nitriding process for construction alloy steels and low-carbon non-alloy steel at a temperature of 560 °C, a pressure of 0.0026 MPa and a duration of up to 6 h. Both authors derived from their experiments an increase in the nitrogen concentration at the steel surface together with a high diffusion rate. For low-pressure nitriding process, these findings are explained by the surface phenomena adsorption, dissociation and desorption [6].

In the case of stainless-steel treatments, several process adjustments are required to maintain corrosion resistance. To prevent chromium depletion the process, temperature is subjected to a limitation. By using a low-temperature treatment, the diffusion depth is limited to a few microns, whereby high temperature treatment results in recrystallization. Furthermore, the passivation layer acts as a diffusion barrier, thus an activation step is necessary. This causes the risk of an inhomogeneous diffusion layer growth by incompleteness of surface activation [3].

In general, thermochemical treatments are applied to bulk materials. Nevertheless, several researchers have combined thermal spray processes and thermochemical treatment and proved the feasibility of stainless-steel coatings using various industrial surface hardening processes [4,7–17]. Plasma, molten metal and salt bath processes show comparable results to bulk material treatments of the same steel type [7,9–14,17]. Conversely, gas nitriding of thermal sprayed coatings, as described in [4,8,15,16], improves the diffusion depth of the enrichment media. Nestler and Lindner, who conducted a gas nitriding process with stainless steel high-velocity oxy-fuel (HVOF) coatings without a surface activation step, explained this finding by the characteristic open porosity of thermal sprayed coatings. The effect is shown for conventional treatment at temperatures above 500 °C, as well as low-temperature processes [4,8,15].

The present study focuses on the modification of a classical gas nitriding process for AISI 316L HVOF thermal-sprayed coatings. The modification comprises a gas pressure modulation and a controlled process gas regime. The gas pressure modulation intends to utilize the open porosity effect of thermal sprayed coatings to increase the nitrogen diffusion depth. An increase of the nitrogen supply at the coating surface is aimed with a controlled nitriding. The temperature-time regime of the gas nitriding process is kept constant at a temperature of 420 °C and a duration of 10 h. The nitriding depth, phase composition, hardness and the wear resistance of the sprayed coatings are compared considering the different nitriding process regimes.

2. Materials and Methods

The AISI 316L coatings were produced using an HVOF K2 system (GTV GmbH, Luckenbach, Germany) with the parameters given in Table 1. The coating was deposited on Ø 40 × 8 mm steel samples of the same grade. Prior to the coating's deposition, the samples were grit-blasted with EK-F24 (Treibacher Industrie AG, Althofen, Austria), a pressure of 3 bar and a distance of 150 mm, under an angle of 70°, then ultrasonically cleaned for 5 min. The coating material was a gas-atomized powder with a particle size fraction of −53 + 20 µm (80.46.1, GTV GmbH, Luckenbach, Germany). After coating production, the samples were ground and polished up to mesh 1000 in order to examine the surface properties after nitriding and the wear tests. Thereby, the coating thickness averaged about 270 µm.

Table 1. Setting parameter of AISI 316L coatings for the HVOF K2 system.

Kerosene [l/h]	Oxygen [l/min]	λ	Nozzle [mm/mm]	Powder Feed Rate [g/min]	Carrier Gas [l/min]	Spray Distance [mm]	Step Size [mm]	Surface Velocity [m/s]
24	900	1.1	150/14	70	2 × 8	350	5	1

Gas nitriding was performed at 420 °C for 10 h in an industrial vacuum chamber retort heat-treatment furnace (WMU Wärmebehandlungsanlagen, Bönen, Germany) equipped with an ammonia cracker (O-SG-9/5, KGO GmbH, Wetter, Germany) and a hydrogen sensor (STANGE

Elektronik GmbH, Gummersbach, Germany) to determine the nitriding potential. The temperature and duration were held constant for all trials. During the trials, the process regime was changed firstly with an industrial process (420 °C/10 h), secondly by adding a pressure modulation (420 °C/10 h PM) and thirdly by performing a pressure modulation in conjunction with a controlled process regime (420 °C/10 h PM + C, Table 2). Ammonia NH_3 with a volume flow of 1000 l/h was used as process gas and mixed with dissociated ammonia in the controlled regime. The amount of cracked ammonia was adjusted according the predetermined nitriding potential maintaining the batch volume. The hydrogen amount was approximately five to six times higher in the controlled regime. The excess pressure in the chamber was varied between 2 mbar and a maximum operation pressure of 50 mbar in a cycle time as short as possible (<10 min). A surface activation step was not considered in the preparation of the gas nitrided samples.

Table 2. Parameter adjustments during gas nitriding trials.

Trial	Process Regime	Pressure Modulation (PM) Over Pressure/Cycle Time	Controlled Process (C)
1	420 °C/10 h	2 mbar	NH_3
2	420 °C/10 h PM	2–50 mbar/<10 min	NH_3
3	420 °C/10 h PM + C	2–50 mbar/<10 min	$NH_3 + N_2 + H_2$

The microstructural characterisation of the nitrided samples included the preparation of cross-sections by hot embedding, grinding and polishing. The cross-sections were wet-etched with the colour etchant Beraha II immediately after polishing. The etching duration varied between 10 and 15 s according to the colour change on the cross-sections. In contrast with the austenitic phase, the supersaturated matrix showed no colouring. Images were recorded using an optical microscope GX51 (Olympus, Shinjuku, Japan) equipped with a SC50 camera (Olympus, Shinjuku, Japan). The nano-indentation was applied at the etched cross-section using a UNAT nano-indenter (ASMEC GmbH, Radeberg, Germany) with a Berkovich tip. In order to determine the hardness values of the different phases, quasi-static measurements were performed with a load of 10 mN and at least 15 repetitions based on DIN EN ISO 14577-1 [18]. The solid solution of nitrogen in the face-centered cubic lattice was investigated by X-ray diffraction (XRD). A D8 DISCOVER diffractometer (Bruker AXS, Billerica, MA, USA) operating with Co Kα radiation (U: 40 kV; I: 40 mA) was used to measure in a diffraction angle (2θ) range from 20° to 130° with a step size of 0.01° and 1.5 s/step. Due to the use of a 1D Lynxeye XE detector (Bruker AXS, Billerica, MA, USA), this corresponded to 288 s/step.

Wear tests were conducted to verify the success of the thermochemical post-treatment. In comparison to the coating in the as-sprayed condition, the gas nitrided samples were tested in ball-on-disk and reciprocating ball-on-plane tests. The ball-on-disk test was carried out with Tetra Basalt Tester (Tetra GmbH, Ilmenau, Germany) based on ASTM G 99 [19] as a dry sliding system and the reciprocating ball-on-plane test was performed with a Wazau SVT 40 device (Wazau GmbH, Berlin, Germany) based on ASTM G 133 [20] as a dry couple. Parameters are given in Table 3. After the tribological testing, the wear tracks were evaluated with contact stylus instrument Hommel Etamic T8000 (Jenoptik GmbH, Villingen-Schwenningen, Germany) to determine the wear area after ball-on-disk testing. For the other test, wear volume was measured with an optical 3D profilometer MikroCAD (LMI Technologies Inc., Burnaby, Canada). The wear tracks were analysed using a scanning electron microscope (SEM) LEO 1455VP (Zeiss, Jena, Germany).

Table 3. Setting parameters for the wear tests.

Ball-on-Disk Test		Reciprocating Ball-on-Plane Test	
Normal load [N]	20	Normal load [N]	26
Radius [mm]	5	Frequency [Hz]	40
Speed [rpm]	96	Time [s]	900
Cycles	15,916	Amplitude [mm]	0.5
Ø Al_2O_3 [mm]	6	Ø Al_2O_3 [mm]	10

3. Results

3.1. Microstructural Analysis

A low-temperature gas nitriding process at 420 °C for 10 h led to nitrogen enrichment in AISI 316L HVOF sprayed coatings without an initial activation step. Figures 1 and 2 illustrate the nitrided coatings depending on the applied process regime. The white layer in the Beraha II-etched cross-sections refers to the S-phase (S), whereas the blue and brown areas represent the initial austenite phase (A). The different colouration of the austenite phase is the result of the etching duration and the preparation of the etchant agent applied for each sample.

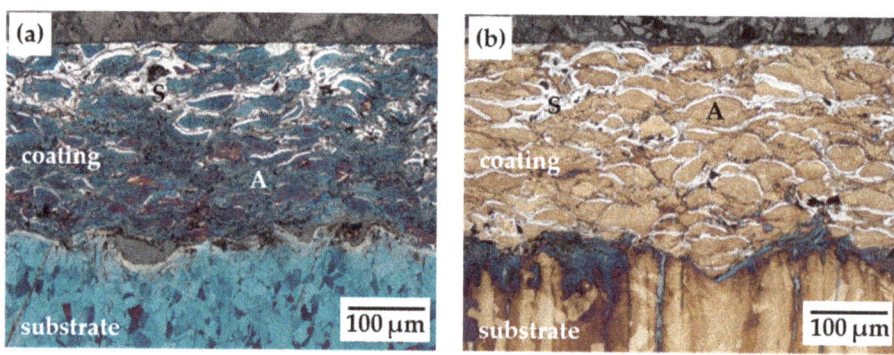

Figure 1. Cross-sectional micrographs of AISI 316L HVOF coating after gas nitriding at (**a**) 420 °C/10 h and (**b**) 420 °C/10 h PM. (S: S-phase, A: austenitic phase).

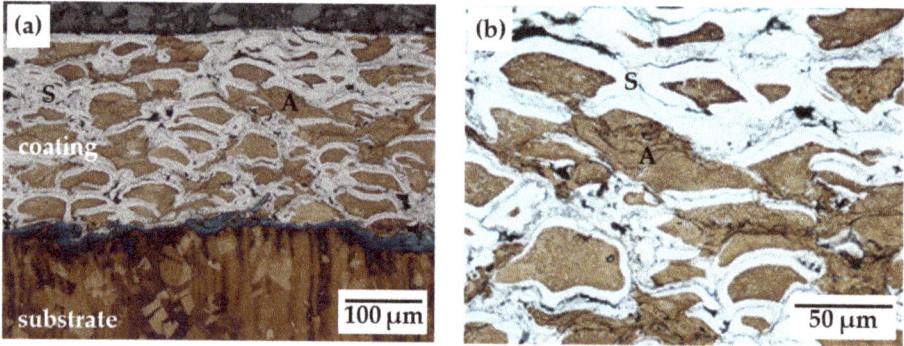

Figure 2. Cross-sectional micrographs of AISI 316L HVOF coating after gas nitriding at 420 °C/10 h with PM + C: (**a**) overview; (**b**) detailed view of the coating. (S: S-phase, A: austenitic phase.)

The formation of the S-phase was inhomogeneous along the coating thickness, starting at the single spray particle's edge. This confirms the gas permeability of the porous thermal spray coating's microstructure. In the as-sprayed condition, the coatings exhibited a porosity of 1.6%. A detailed description of the coating's microstructure before nitriding is given in [8]. The penetration depth was enhanced by a pressure modulation of the gas donator (Figure 1b). The gas exchange improved through the coating's open porosity. In contrast, a plasma thermochemical treatment [7,9,10,12–14] or precipitation hardening [4,8] of thermal spray steel coatings generated a homogenous S-phase or compound layer up to 20 µm or above 100 µm in depth, respectively.

Significant improvements can be achieved by using dissociated ammonia for the same process duration. The diffusion zone increased at the spray particle's edge and a uniform diffusion depth was reached up to the substrate surface (Figure 2a). Small particles were nitrided completely (Figure 2b). The coatings exhibited a similar S-phase fraction in comparison with results for a 30-h duration without a controlled gas regime [8]. Hence, a notable time reduction of the process was realised. Consequently, the use of dissociated ammonia under modulated gas pressure ensured a high nitrogen supply within the coating's structure. As a result, the diffusion in depth and a homogenous distribution of the austenite and S-phase were improved. These results are in accordance with effects recognized for high- and low-pressure gas nitriding of bulk materials under similar process conditions [5,6].

Figure 3 shows the XRD patterns of the HVOF coatings considering the different treatment states in conjunction with the as-sprayed condition. The untreated coating exhibited characteristic peaks of the austenite phase and additional minor peaks that corresponded to the ferrite phase. Depending on the setting of the gas nitriding process, the peak intensity decreased. Additional peaks of the expanded austenite appeared at lower angles compared to the initial austenitic phase. The peaks shifted and intensity increased with the amount of the S-phase fraction. A higher magnification of the peak shift for the lattice planes {111} and {200} is illustrated in Figure 3b. The broad S-phase peak indicates a superposition of different lattice expansions and an inhomogeneous interstitial dissolution of nitrogen within the coating's microstructure. Higher lattice expansion equates a higher nitrogen enrichment, as observed for the pressure-modulated and controlled gas nitriding. The results differ from the XRD pattern of a low-temperature thermochemical-treated bulk material and a plasma thermochemical-treated thermal spray AISI 316L coating. These revealed clear peak shifts of the austenite lattice planes, indicating the S-phase [2,7,15]. Nitride phases like CrN or Fe_4N can be excluded by the XRD measurements.

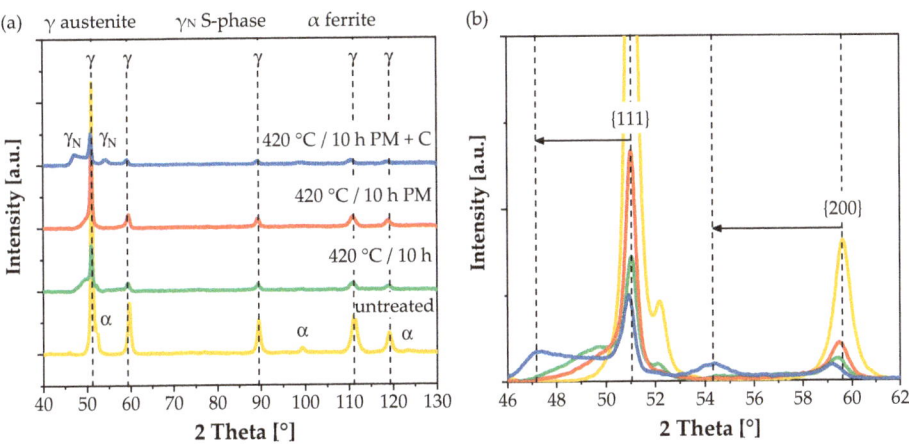

Figure 3. XRD diagrams of AISI 316L HVOF coatings before and after gas nitriding: (**a**) overview; (**b**) detailed view of the peak shift of {111} and {200} lattice planes.

3.2. Hardness and Wear Resistance

The coating's cross-sections showed a two-phase microstructure after gas nitriding. Table 4 summarizes the nano-hardness of the different microstructure domains. The austenitic phase of the nitrided samples exhibited a nano-hardness of 440 HV_{10mN} on average. In comparison with the untreated state, the increase in hardness was affected by the nearby S-phase. The S-phase was more than twice as hard and ranged between 870 and 1000 HV_{10mN}. The hardness values are in accordance with the results of plasma nitrided cold sprayed AISI 316L coatings [13]. In general, a strong shift of the diffraction peak corresponded to high nano-hardness values. This relationship was in particular valid for greater portion of S-phase fractions. All samples showed a superposition of the dissolved nitrogen lattice expansion. A decreasing nitrogen concentration from the enriched spray particles' edges was reasonable for a broad peak appearance. Hence, a gradation in hardness can be assumed.

Table 4. Hardness of the AISI 316L HVOF coatings before and after gas nitriding.

State of Treatment	Untreated	420 °C/10 h	420 °C/10 h PM	420 °C/10 h PM + C
Austenitic phase HV_{10mN}	316 ± 59	462 ± 77	420 ± 39	449 ± 41
S-phase HV_{10mN}		874 ± 136	1005 ± 65	971 ± 68

Low-temperature gas nitriding improved the wear resistance of the coatings in the ball-on-disk and reciprocating ball-on-plane test conditions (Figure 4). The wear rate was deeply influenced by the wear mechanism. Because of the high deformation of the austenite phase, the untreated AISI 316L steel coating showed severe wear in both wear tests (Figure 5a). Adhesive wear prevailed also in the gas nitrided samples at 420 °C/10 h, which showed a significant decrease in wear area and volume, respectively (Figure 5b). In addition, particle breakouts increased in the reciprocating ball-on-plane test due to the frequent contact of the surface with the counterbody. In comparison with this, the wear of the pressure-modulated sample increased. The higher nano-hardness values resulted in a certain change in wear mechanism. Figure 5c shows a predominately abrasive wear with deep grooves in the direction of sliding for the ball-on-disk test. These were caused by the breakout of hardened spray particles acting as abrasives in combination with inhomogeneous nitriding at the surface (Figure 1b). However, the adhesive wear dominated in the reciprocated ball-on-plane test.

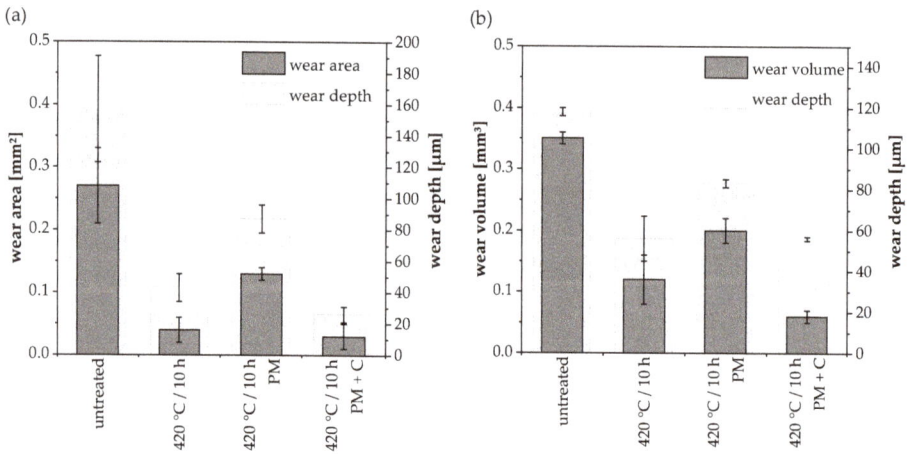

Figure 4. Results of the wear tests of AISI 316L coatings before and after gas nitriding, (a) ball-on-disk test and (b) reciprocating ball-on-plane test.

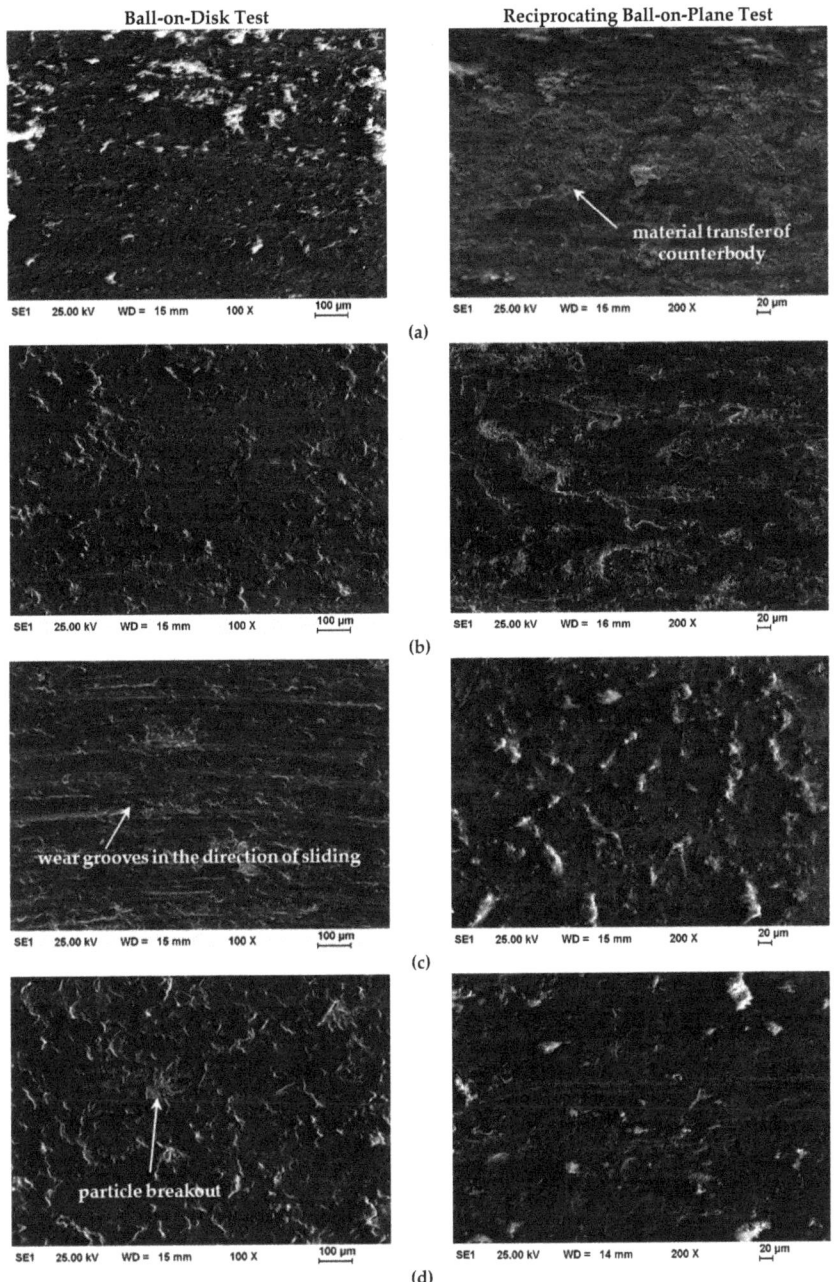

Figure 5. SEM micrographs of the wear tracks of AISI 316L coatings before and after gas nitriding after the ball-on-disk and reciprocating ball-on-plane tests. (**a**) AISI 316L HVOF coating in untreated condition; (**b**) AISI 316L HVOF coatings after gas nitriding at 420 °C/10 h; (**c**) AISI 316L HVOF coatings after gas nitriding at 420 °C/10 h PM; (**d**) AISI 316L HVOF coatings after gas nitriding at 420 °C/10 h PM + C.

The highest wear resistance was proven for the gas nitrided samples with dissociated ammonia in the pressure-modulated regime (Figure 4) possessing a homogenous nitrogen enrichment and S-phase fraction. The main wear mechanism can be assigned to adhesive wear indicated by plastic deformation and minor particle breakouts in both tests (Figure 5d). A reduction of approximately 89% in wear area for the ball-on-disk test condition and 83% in wear volume for the reciprocated ball-on-plane test condition can be achieved by thermochemical treatment of coatings. In conclusion, a post gas nitriding of AISI 316L HVOF coatings led to a significant wear improvement with a general better performance in the ball-on-disk test.

4. Conclusions

The microstructural evolution in HVOF-sprayed AISI 316L coatings during gas nitriding was studied with respect to parameter settings. A pressure modulation within the post-treatment step improved the penetration depth. It was found that the diffusion depth was decisively increased by using dissociated ammonia. In addition to this, the process efficiency related to the processing time was improved. A homogeneous diffusion layer growth within the single spray particles from the coatings' surfaces to the substrate was proven for an adjusted gas donator composition and a pressure-modulated process management. This can be explained by a higher activity and constant exchange with renewing of the donator media. The expanded austenite showed significantly increased hardness in comparison with the initial austenitic phase. Gas nitriding of AISI 316L HVOF coatings resulted in a significant enhanced wear resistance. Due to the thermochemical treatment, a change in wear mechanism could be recognized. Abrasive wear was indicated by grooves within the austenitic phase. These were caused by worn out particles of the hard S-phase. In the case of a homogenous hardened surface, the adhesive wear occurred, resulting in best wear resistance shown by a combined pressure-modulated and controlled gas nitriding process. From the results of the present study, the general feasibility of the novel processing approach is confirmed. The constant hardening depth within the single spray particles allows a reliable prediction of the coating properties regardless surface distance.

Author Contributions: P.K. and T.L. (Thomas Lindner) conceived and designed the experiments. P.K. performed the experiments, analysed the data and wrote the paper. K.B. and U.R. adapted and executed the gas nitriding process. T.L. (Thomas Lampke) directed the research and contributed to the discussion and interpretation of the results.

Funding: The authors gratefully acknowledge the Arbeitsgemeinschaft Industrieller Forschungsvereinigungen "Otto von Guericke" e.V. (AiF) for support of this work (AiF-No. KF2152613WZ4 & KF2550004WZ4) with funds from the German Federal Ministry for Economic Affairs and Energy. The publication costs of this article were funded by the German Research Foundation/DFG-392676956 and the Technische Universität Chemnitz in the funding programme Open Access Publishing.

Acknowledgments: The authors thank Thomas Mehner for conducting the XRD measurements, Christel Pönitz and Paul Clauß for support in metallographic investigation. Special thanks is dedicated to the CEO Ulrich Reese who strongly promoted the investigations of thermally sprayed coatings on an industrial scale. We deeply regret that he passed away right before the officially acceptance for publication of this joined paper.

Conflicts of Interest: The authors declare no conflict of interest.

References

1. Christiansen, T.L.; Somers, M.A.J. Low-temperature gaseous surface hardening of stainless steel: The current status. *Int. Mater. Res.* **2009**, *100*, 1361–1377. [CrossRef]
2. Bell, T. Current status of supersaturated surface engineering S-phase materials. *Key Eng. Mater.* **2008**, *373–374*, 289–295. [CrossRef]
3. Bottoli, F.; Jellesen, M.S.; Christiansen, T.L.; Winther, G.; Somers, M.A.J. High temperature solution-nitriding and low-temperature nitriding of AISI 316: Effect on pitting potential and crevice corrosion performance. *Appl. Surf. Sci.* **2018**, *431*, 24–31. [CrossRef]
4. Nestler, M.C.; Spies, H.; Hermann, K. Production of duplex coatings by thermal spraying and nitriding. *Surf. Eng.* **1996**, *12*, 299–302. [CrossRef]

5. Alekseeva, M.S.; Gress, M.A.; Scherbakov, S.P.; Gerasimov, S.A.; Kuksenova, L.I. The influence of high-pressure gas nitriding on the properties of martensitic steels. *Met. Sci. Heat Treat.* **2017**, *59*, 524–528. [CrossRef]
6. Wolowiec-Korecka, E.; Michalski, J.; Kucharska, B. Kinetic aspects of low-pressure nitriding process. *Vacuum* **2018**, *155*, 292–299. [CrossRef]
7. Adachi, S.; Ueda, N. Wear and corrosions properties of cold-sprayed AISI 316L coatings treated by combined plasma carburizing and nitriding at low temperature. *Coatings* **2018**, *8*, 456. [CrossRef]
8. Lindner, T.; Kutschmann, P.; Löbel, M.; Lampke, T. Hardening of HVOF-sprayed austenitic stainless-steel coatings by gas nitriding. *Coatings* **2018**, *8*, 348. [CrossRef]
9. Adachi, S.; Ueda, N. Combined plasma carburizing and nitriding of sprayed AISI 316L coating for improved wear resistance. *Surf. Coat. Technol.* **2014**, *259*, 44–49. [CrossRef]
10. Park, G.; Bae, G.; Moon, K.; Lee, C. Effect of plasma nitriding and nitrocarburizing on HVOF-sprayed stainless steel coatings. *J. Therm. Spray Technol.* **2013**, *22*, 1366–1373. [CrossRef]
11. Wielage, B.; Rupprecht, C.; Lindner, T.; Hunger, R. Surface modification of austenitic thermal spray coatings by low-temperature carburization. In Proceedings of the International Thermal Spray Conference & Exposition, Long Beach, CA, USA, 11–14 May 2015.
12. Adachi, S.; Ueda, N. Formation of S-phase layer on plasma sprayed AISI 316L stainless steel coating by plasma nitriding at low temperature. *Thin Solid Films* **2012**, *523*, 11–14. [CrossRef]
13. Adachi, S.; Ueda, N. Formation of expanded austenite on a cold-sprayed AISI 316L coating by low-temperature plasma nitriding. *J. Therm. Spray Technol.* **2015**, *24*, 1399–1407. [CrossRef]
14. Lindner, T.; Löbel, M.; Lampke, T. Phase Stability and Microstructure Evolution of Solution-Hardened 316L Powder Feedstock for Thermal Spraying. *Metals* **2018**, *8*, 1063. [CrossRef]
15. Lindner, T.; Mehner, T.; Lampke, T. Surface modification of austenitic thermal-spray coatings by low-temperature nitrocarburizing. *IOP Conf. Ser. Mater. Sci. Eng.* **2016**, *118*. [CrossRef]
16. Piao, Z.-Y.; Xu, B.S.; Wang, H.D.; Wen, D.H. Influence of surface nitriding treatment on rolling contact behavior of Fe-based plasma sprayed coating. *Appl. Surf. Sci.* **2013**, *266*, 420–425. [CrossRef]
17. Mindivan, H. Investigating tribological charateristics of HVOF sprayed AISI 316 stainless steel coating by pulsed plasma nitriding. *IOP Conf. Ser. Mater. Sci. Eng.* **2018**, *295*. [CrossRef]
18. ISO. *DIN EN ISO 14577-1: Metallic Materials-Instrumented Indentation Test for Hardness and Materials Parameters—Part 1: Test Method (ISO 14577-1:2015)*; International Organization for Standardization: Geneva, Switzerland, 2018; German version: EN ISO 14577-1:2015.
19. ASTM International. *ASTM G 99 Standard Test Method for Wear Testing with a Pin-on-Disk Apparatus*; ASTM International: West Conshohocken, PA, USA, 2016.
20. ASTM International. *ASTM G 133 Standard Test Method for Linearly Reciprocating Ball-on-Flat Sliding Wear*; ASTM International: West Conshohocken, PA, USA, 2016.

© 2019 by the authors. Licensee MDPI, Basel, Switzerland. This article is an open access article distributed under the terms and conditions of the Creative Commons Attribution (CC BY) license (http://creativecommons.org/licenses/by/4.0/).

Article

Durability of Gadolinium Zirconate/YSZ Double-Layered Thermal Barrier Coatings under Different Thermal Cyclic Test Conditions

Satyapal Mahade [1],*, Nicholas Curry [2], Stefan Björklund [1], Nicolaie Markocsan [1] and Shrikant Joshi [1]

1. Department of Engineering Science, University West, 46186 Trollhattan, Sweden
2. Treibacher Industrie AG, Auer von Welsbachstr, 1, A-9330 Althofen, Austria
* Correspondence: satyapal.mahade@hv.se

Received: 20 May 2019; Accepted: 8 July 2019; Published: 11 July 2019

Abstract: Higher durability in thermal barrier coatings (TBCs) is constantly sought to enhance the service life of gas turbine engine components such as blades and vanes. In this study, three double layered gadolinium zirconate (GZ)-on-yttria stabilized zirconia (YSZ) TBC variants with varying individual layer thickness but identical total thickness produced by suspension plasma spray (SPS) process were evaluated. The objective was to investigate the role of YSZ layer thickness on the durability of GZ/YSZ double-layered TBCs under different thermal cyclic test conditions i.e., thermal cyclic fatigue (TCF) at 1100 °C and a burner rig test (BRT) at a surface temperature of 1400 °C, respectively. Microstructural characterization was performed using SEM (Scanning Electron Microscopy) and porosity content was measured using image analysis technique. Results reveal that the durability of double-layered TBCs decreased with YSZ thickness under both TCF and BRT test conditions. The TBCs were analyzed by SEM to investigate microstructural evolution as well as failure modes during TCF and BRT test conditions. It was observed that the failure modes varied with test conditions, with all the three double-layered TBC variants showing failure in the TGO (thermally grown oxide) during the TCF test and in the ceramic GZ top coat close to the GZ/YSZ interface during BRT. Furthermore, porosity analysis of the as-sprayed and TCF failed TBCs revealed differences in sintering behavior for GZ and YSZ. The findings from this work provide new insights into the mechanisms responsible for failure of SPS processed double-layered TBCs under different thermal cyclic test conditions.

Keywords: double-layered TBC; gadolinium zirconate; suspension plasma spray; thermal cyclic fatigue; burner rig test; yttria stabilized zirconia

1. Introduction

Thermal barrier coatings (TBCs) enhance the efficiency of a gas turbine engine by allowing them to operate at higher temperatures, in order to lower engine emissions and improve fuel economy [1]. At higher operating temperatures (>1200 °C), the existing state-of-the-art top coat TBC candidate, 7–8 wt. % yttria stabilized zirconia (YSZ), has several limitations such as phase instability, high sintering rates, etc. [2,3]. Another major drawback of YSZ as a top coat material is also its susceptibility to CMAS infiltration above 1200 °C, which limits TBC longevity [4–7].

Alternative ceramic top coat materials for TBC application such as gadolinium zirconate (GZ), have been shown to possess lower thermal conductivity and excellent phase stability compared to YSZ at high temperatures [8]. The other advantage of GZ over YSZ is its excellent CMAS (Calcium-Magnesium-Alumino-Silicates) attack resistance [9]. However, GZ has drawbacks such as inferior fracture toughness and poor thermochemical compatibility with alumina (thermally grown oxide), which limits its durability [10,11]. To overcome these drawbacks, double-layered TBCs with GZ as the top layer and YSZ as the base layer have been used [12,13]. A double-layered, GZ/YSZ TBC design exploits the merits of YSZ (high fracture toughness close to the TGO/bond coat) and GZ (excellent CMAS resistance, low thermal conductivity etc.). Furthermore, the GZ/YSZ multi-layered TBCs were shown to be more durable compared to a YSZ single layer TBC [12]. Failure analysis of the GZ/YSZ multi-layered TBCs investigated so far with a layer thickness ratio of 4:1 for GZ and YSZ revealed spallation of the GZ layer close to its interface with YSZ and at a distance of less than 100 µm from the bond coat [12,14]. The reason for this was attributed to lower fracture toughness of GZ than YSZ [15]. Fracture toughness plays a crucial role in governing spallation in the ceramic as it resists crack propagation due to accumulation of stresses in the coating from various sources (coefficient of thermal expansion [CTE] mismatch between ceramic and metallic substrate, oxidation of bond coat, etc.) during thermal cycling [8]. The specific purpose of this study was to assess if the durability of GZ/YSZ double-layered TBC could be improved by moving the inferior fracture toughness material, i.e., GZ, further away from the bond coat by increasing the underlying YSZ layer thickness. By doing so, the relatively higher fracture toughness material (YSZ) is presented at the probable failure location, which could lead to improved durability by delaying the onset of failure. Durability of TBCs and their associated failure mechanisms vary with the test conditions employed (such as cycling conditions, exposure temperature and time etc.) [16]. Furthermore, the choice of substrate (composition) has an influence on the durability as it leads to differential CTE mismatch with the top coat, resulting in differing stress state in the TBC [17,18]. Previously, burner rig test comprising of short exposure time (75 s of heating cycle) and relatively lower exposure temperature (1350 °C), were employed to evaluate the performance of GZ/YSZ double-layered TBCs deposited on Hastelloy-X substrates [19].

In this work, three double-layered GZ/YSZ TBCs with variable YSZ thickness i.e., 400GZ/100YSZ, 250GZ/250YSZ and 100GZ/400YSZ, where the prefixed numbers represent layer thickness in µm, were deposited by SPS process. SPS process was opted over other TBC deposition techniques such as EB-PVD (Electron Beam- Physical Vapor Deposition) or APS (Atmospheric Plasma Spray) due to its capability to produce columnar microstructured TBCs via plasma route. Columnar microstructured TBCs are desirable for higher cyclic lifetime and higher erosion resistance [20,21]. The GZ/YSZ TBCs were subjected to two different thermal cyclic test conditions, one without a thermal gradient (thermal cyclic fatigue or TCF) and the other with a thermal gradient (burner rig test or BRT) across the TBC test specimen. For BRT, TBCs were deposited on Inconel-738 substrates and exposed to a relatively higher surface temperature and exposure time than the ones reported elsewhere [19]. The BRT and TCF failed TBC specimens were analyzed by SEM to gain insights into the mechanisms responsible for failure.

2. Experimental Details

Two different substrates, namely Hastelloy-X and Inconel-738, with dimensions 50 mm × 30 mm × 6 mm and 30 mm dia × 3 mm thickness were used for the TCF and BRT tests, respectively. The substrates were grit blasted to create a surface roughness of approximately 3 µm Ra. A NiCoCrAlY bond coat (AMDRY 386, Oerlikon Metco, Westbury, New York, NY, USA) was first deposited on the surface using High Velocity Air Fuel (HVAF) process (M3 gun, UniqueCoat, Oilville, Virginia, VA, USA). The thickness of bond coat was kept at 190 µm ± 10 µm in all the investigated specimens. HVAF sprayed bond coat may contain unmelted droplets and loosely bound particles on the surface which could hinder mechanical adhesion of the TBC. Therefore, grit blasting of the bond coated surface was carried out to get rid of the loosely bound particles and create a surface roughness of approximately 5 µm Ra. After grit blasting, the surface was cleaned with pressurized air to remove loosely bound grit particles.

Ethanol based, commercial 8YSZ and GZ suspensions supplied by Treibacher Industrie AG, Althofen, Austria were used. Both the 8YSZ suspension (AuerCoat YSZ) and the GZ suspension (AuerCoat Gd-Zr) comprised of powders with a median size of 500 nm and a solid load content of 25 wt. %.

The bond coated substrates were preheated prior to top coat deposition in order to remove volatile impurities from the surface. An axial-feed capable plasma torch (Axial III, Mettech, Vancouver, BC, Canada) was used to deposit the ceramic layers. Identical spray parameters (Table 1) were used for deposition of YSZ and GZ layers. The first double layered TBC variation had 400 µm thick GZ top layer and 100 µm thick YSZ base layer, which is denoted as 400GZ/100YSZ. Similarly, the second and third double-layered TBC variations are denoted as 250GZ/250YSZ and 100GZ/400YSZ.

Table 1. Suspension plasma spray parameters for YSZ and GZ layers in the three double layered TBC variants.

Parameters	YSZ	GZ
Solid load content (wt. %)	25	25
Median particle size (nm)	550	550
Solvent	Ethanol	Ethanol
Stand off distance (mm)	100	100
Enthalpy (kJ/l)	12.5	12.5
Atomizing gas flow (L/min)	20	20

Microstructure of the as-sprayed TBCs was characterized using a scanning electron microscope (HITACHI TM 3000, Hitachi High-Technologies Corporation, Tokyo, Japan). The porosity content in GZ and YSZ layers of as-sprayed and TCF failed TBCs were analyzed using an open software called 'ImageJ' (version 1.52p, University of Wisconsin, Wisconsin, US) [22] by considering twenty five different cross-sectional SEM micrographs at high magnification (5000×). Furthermore, column gaps in the TBC were not considered during porosity measurement. For simplicity, one TBC variation (250GZ/250YSZ) was chosen for the porosity measurement of GZ and YSZ layers in the as-sprayed condition since all the investigated TBCs were deposited using identical spray parameters. After TCF test, porosity evolution in GZ and YSZ layers for all three TBC variations was measured and compared with as-sprayed condition.

In the TCF test, the TBC specimens were exposed to 1100 °C for 1 hr and were later, cooled to 100 °C in 10 min using compressed air. This cycle was repeated until specimen failure, which was deemed to be 20% visible TBC spallation. After the completion of each cycle (heating and cooling), photograph of the specimens were taken by integrating a high-resolution camera with LabVIEW software in order to monitor the progress of failure. The temperature during the heating cycle was monitored using two thermocouples, which were placed at different locations in the heating chamber. Further details regarding the TCF test setup are disclosed in our previous work [23]. It should be noted that the TCF test was conducted in the absence of a thermal gradient across the TBC specimen. Three specimens of each coating variation were subjected to a TCF test and the mean value and standard deviation are reported.

In the burner rig tests (BRT), performed at Forschungszentrum Jülich, Germany, the TBC surface was exposed to a surface temperature of 1400 °C with the rear of the specimen maintained at a temperature of 1050 °C. The test specimen was heated periodically using natural gas/oxygen burners whereas the backside of the specimen was cooled using compressed air. The surface temperature was measured using a land infrared pyrometer whereas the backside (substrate) temperature was measured using a NiCr/Ni thermocouple. Each cycle involved 5 min. heating followed by 2 min of cooling. Failure criteria was 30% visual spallation. Further details regarding the test setup and TBC specimen geometry are disclosed elsewhere [24]. Two specimens were tested for each TBC variant and their mean values are reported.

3. Results and Discussion

3.1. Microstructure

The cross sectional SEM micrograph of the 400GZ/100YSZ double-layered TBC showed columnar microstructure, according to Figure 1a. A similar columnar microstructure of the GZ/YSZ double layered TBC processed by SPS process was reported elsewhere [25]. The thickness of top GZ layer was measured to be approximately 410 μm and the base YSZ layer was approximately 95 μm. The GZ/YSZ interface in this case seems to be continuous and free from delamination cracks, according to Figure 1b. This is highly desirable since any cracking or discontinuity at the interface can promote delamination and adversely influence durability. Similarly, the cross sectional SEM micrographs of 250GZ/250YSZ and 100GZ/400YSZ TBCs showed a columnar microstructure along with a delamination free GZ/YSZ interface, according to Figure 1c–f, respectively.

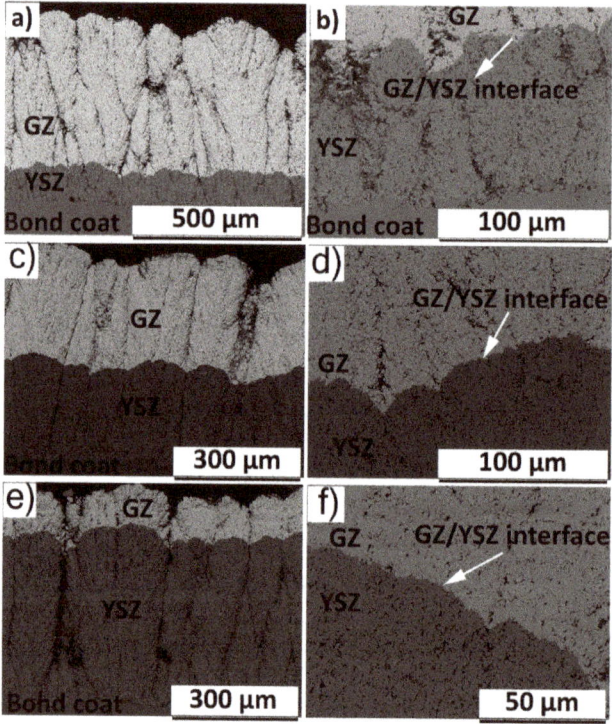

Figure 1. SEM (scanning electron microscope) micrograph of as-sprayed TBCs: (**a**) 400GZ/100YSZ cross section; (**b**) GZ/YSZ interface of 400GZ/100YSZ; (**c**) 250GZ/250YSZ cross section; (**d**) GZ/YSZ interface of 250GZ/250YSZ; (**e**) 100GZ/400YSZ cross section; (**f**) GZ/YSZ interface of 100GZ/400YSZ.

3.2. Porosity

Porosity content in the individual layers comprising the TBC specimen, i.e., GZ and YSZ, were measured in as-sprayed condition for 250GZ/250YSZ. The YSZ layer showed a higher porosity content than the GZ layer, although the spray parameters and suspension properties were kept the same, see Table 1 and Figure 2. It should be noted that GZ has a lower melting temperature (2570 °C) than YSZ (2700 °C). Therefore, the GZ splats undergo a greater degree of melting than YSZ. According to the understood theory of plasma spraying, this should result in a relatively denser coating for GZ than for YSZ. This explains the higher porosity content observed for YSZ than for GZ. Similar findings of

YSZ showing higher porosity compared to GZ were reported in the past under similar processing conditions [25].

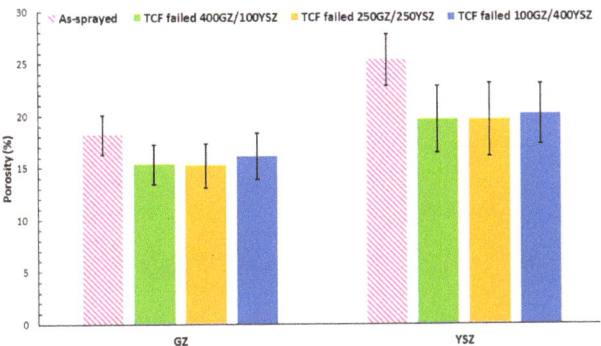

Figure 2. Porosity content of as-sprayed GZ and YSZ layers and after failure.

The porosity content in a TBC can be expected to progressively change with time when subjected to a burner rig test as a result of sintering. Moreover, the porosity can plausibly also vary along the TBC thickness due to the presence of a thermal gradient across the test specimen, leading to greater sintering in the ceramic layer closer to the burner. Therefore, it is a challenge to compare the porosity content in a TBC before and after exposure to BRT. However, in a TCF test, the lack of a thermal gradient across the test specimen ensures that the extent of sintering in the ceramic material is independent of thickness across the TBC. Therefore, in this work, porosity evolution in GZ and YSZ layers were compared in as-sprayed and TCF failed condition for 400GZ/100YSZ, 250GZ/250YSZ and 100GZ/400YSZ TBCs. After the TCF test, the individual GZ and YSZ layers showed a reduction in porosity compared to the as-sprayed condition, according to Figure 2. Sintering is known to lead to an increase in stiffness of the TBC, which could result in loss of strain tolerance and eventually lead to TBC failure [26,27]. Therefore, it is desirable to have a ceramic top coat material which can provide better sintering resistance. Furthermore, according to the mean porosity values in Figure 2, the YSZ layer showed higher reduction in porosity than GZ in all the three variants of TCF tested TBCs when compared to the as-sprayed condition. However, the error bar consideration shows no significant difference in the sintering resistance of GZ and YSZ. Cao et al. reported higher sintering resistance of rare earth based pyrochlores than YSZ [28]. The reason for higher sintering resistance of pyrochlores than YSZ was attributed to the fact that the oxygen anion vacancies in pyrochlores are arranged in an orderly fashion compared to YSZ [28].

3.3. TBCs Subjected to TCF Test

In TCF, the test conditions are relatively harsh due to the fact that the top coat, bond coat and the substrate are exposed to the same temperature, i.e., there is no thermal gradient across the TBC test specimen. The TCF test results indicate that the GZ/YSZ double-layered TBC with higher YSZ thickness (100GZ/400YSZ) showed lower durability whereas the TBC with lower YSZ thickness (400GZ/100YSZ) showed higher durability, as depicted in Figure 3. In a TBC, the bond coated surface could be accessed (oxidized) by oxygen via two possible routes i.e., from the open porosity through TBC and the oxygen ion vacancies in the crystal structure of the ceramic. It is speculated that the oxygen penetration resistance of rare earth zirconate based pyrochlores is higher than YSZ due to their cubic crystal structure having a systematic arrangement of the oxygen ion vacancies [28,29]. Moreover, the porosity content in the GZ layer has also been shown to be lower than that in the YSZ layer, thereby further contributing to the better oxygen penetration resistance.

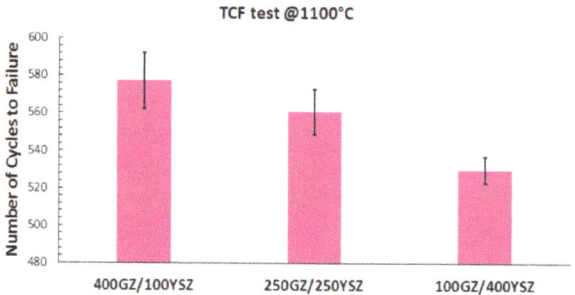

Figure 3. Thermal cyclic fatigue (TCF) life of GZ/YSZ double-layered TBCs.

The cross sectional SEM micrograph of failed 400GZ/100YSZ showed column gap widening in the double-layered TBC, according to Figure 4a. The reason for column gap widening could be attributed to the tensile stresses in the TBC during the heating cycle, where the metallic substrate expands more than the ceramic. The photograph of failed TBC specimen showed TBC spallation and the oxidized bond coat surface being exposed, according to Figure 4b. At high magnification, the failure in the TGO layer due to horizontal crack propagation could be seen, according to Figure 4c. Similar failure mode in the GZ/YSZ double layered TBCs when subjected to TCF test was reported elsewhere [12]. The thermally grown oxide (TGO) layer at failure showed a thickness of approximately 6–7 µm. Lu et al. and Smialek et al. previously reported the critical TGO thickness at failure in TCF tested specimens to be approximately 7–8 µm [30–32]. The failed specimen does not show blue failure, indicating the presence of alumina in the failed region of TGO [33].

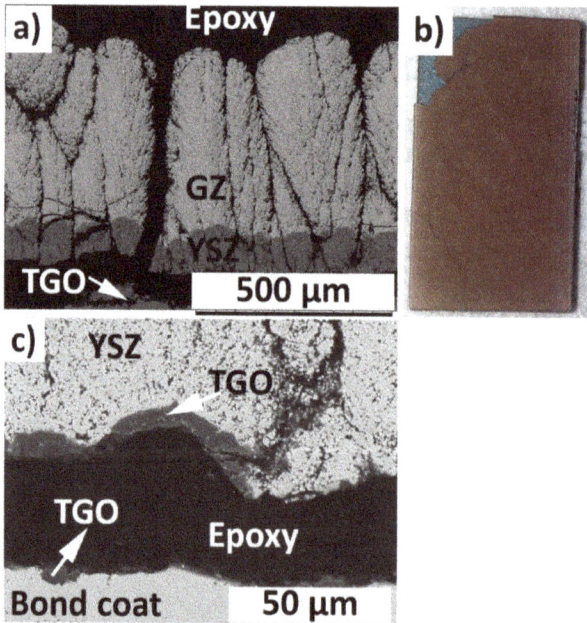

Figure 4. SEM micrograph of TCF (thermal cyclic fatigue) failed (570 cycles) 400GZ/100YSZ (**a**) Cross section and (**b**) Photograph (**c**) TGO (thermally grown oxide).

The cross sectional SEM micrograph of the failed 250GZ/250YSZ and 100GZ/400YSZ double layered TBCs also showed failure in the TGO layer, according to Figures 5a and 6a. The TGO thickness at failure was also measured to be approximately 6–7 µm in these TBC variations. The photographs of the TBC after failure also showed spallation of the ceramic coating from the test specimen edges, according to Figures 5b and 6b. The high magnification cross sectional SEM micrograph of failed TGO layer in the case of 100GZ/400YSZ is shown in Figure 6c.

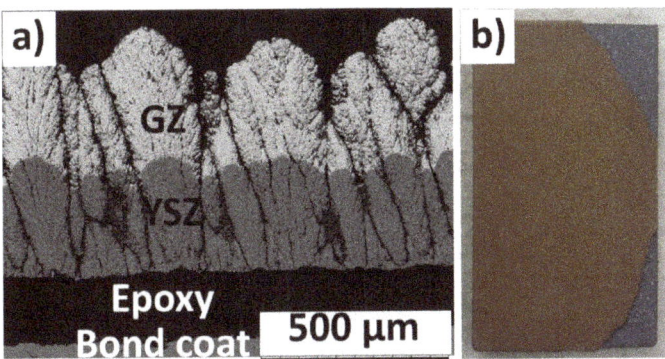

Figure 5. SEM micrograph of TCF failed (552 cycles) 250GZ/250YSZ (**a**) Cross section and (**b**) Photograph.

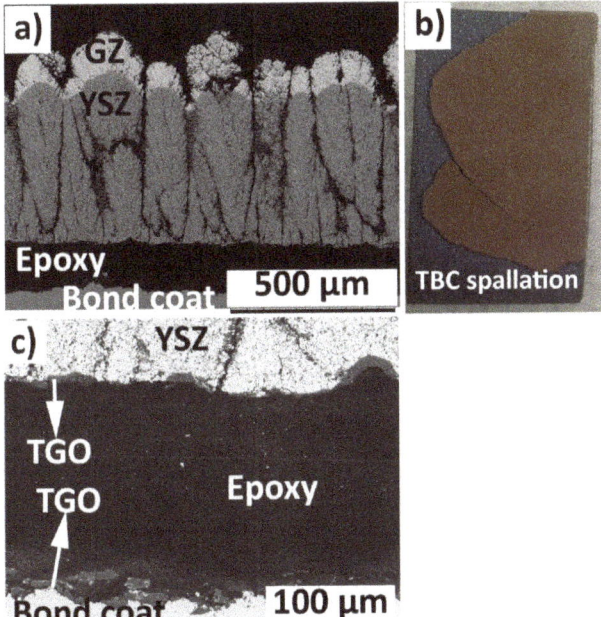

Figure 6. SEM micrograph of TCF failed (528 cycles) 100GZ/400YSZ (**a**) Cross section; (**b**) photograph and (**c**) TGO.

Failure in the double-layered TBC variations (400GZ/100YSZ, 250GZ/250YSZ, 100GZ/400YSZ) when subjected to TCF test appears to be similar, i.e., spallation of TBC due to failure in the TGO. However, their TCF durability differs, with the 400GZ/100YSZ showing the highest durability and

100GZ/400YSZ showing the lowest. Oxidation of bond coat and attaining critical TGO thickness has been previously reported to be the limiting step governing the durability of TBCs under a TCF test, and this has been reaffirmed by the present results.

In the TCF test, the relatively longer exposure to high temperature allows sufficient time for oxidation of the bond coat, while the ceramic coating simultaneously undergoes a degree of sintering. Furthermore, the CTE mismatch between the metallic substrate and ceramic coating during the heating and cooling cycle leads to accumulation of strain energy in the coating. These three mechanisms (oxidation of bond coat, sintering, CTE mismatch) compete for the TBC failure and reaching a critical TGO thickness via bond coat oxidation appears to be the dominant failure mode under TCF test conditions.

3.4. TBCs Subjected to BRT

The TBC surface temperature in BRT test was chosen as 1400 °C in order to replicate the desired service temperature of an aero engine. In BRT, the durability of double-layered TBCs showed a ranking similar to that during TCF testing, where the TBC with higher YSZ thickness (100GZ/400YSZ) in the GZ/YSZ double-layer coating showed lower durability and vice-versa, although the absolute values of the lifetime were different, see Figure 7. Previous findings demonstrated that the absolute values of durability results would improve (longer lifetime) when the TBC surface temperature is lowered [23]. Furthermore, ranking of the coatings, in terms of durability, were shown to be the same with lower exposure temperature [23].

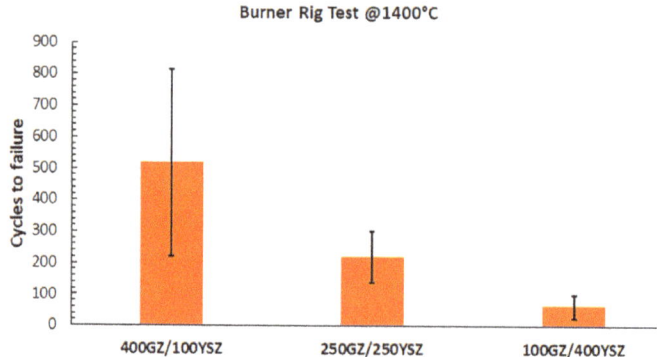

Figure 7. Burner rig test (BRT) life of GZ/YSZ double-layered TBCs.

Failure analysis of the BRT 400GZ/100YSZ TBC showed spallation of top GZ layer from region close to interface with YSZ, according to Figure 8a,c. Exposure of TBCs to thermal cyclic test results in accumulation of strain energy in the ceramic coating. This could be attributed to the coefficient of thermal expansion (CTE) mismatch between the ceramic coating (10.4 × 10^{-6}/K for GZ and 11.5 × 10^{-6}/K at 30–1000 °C [8]) and the metallic substrate (16–17 × 10^{-6}/K at 1000 °C [34]) during the heating cycle (resulting in tensile stresses in the TBC) and cooling cycle (compressive stresses in the TBC). Viswanathan et al. used the concept of available elastic energy to explain the failure modes observed in their findings for APS processed GZ/YSZ TBCs subjected to thermal cyclic test [35]. In their findings, it was reported that TBC failure would occur when the available elastic energy exceeds the critical stress intensity factor (fracture toughness) of the ceramic [35]. YSZ has higher fracture toughness than GZ (approximately double) [15]. Recently, Zhou et al. reported the fracture toughness of SPS processed, porous, columnar microstructured YSZ TBC was approximately 1.0 Mpa.m$^{1/2}$ [36] whereas, for SPS processed porous GZ, fracture toughness was 0.48 Mpa.m$^{1/2}$ [10]. Therefore, in a GZ/YSZ double-layered TBC, it is expected that the GZ layer would allow crack propagation with relative ease compared to YSZ. After reaching a certain number of cycles in the burner rig, the stored elastic strain energy in the GZ/YSZ system presumably exceeds the fracture toughness of GZ and, hence, results in spallation of the GZ layer from a region close to GZ/YSZ interface. Similar horizontal cracks close to GZ/YSZ interface were reported previously elsewhere for GZ/YSZ double-layered TBCs processed by APS [12]. The photograph of failed 400GZ/100YSZ TBC showed failure in the ceramic layer, leaving behind some part of the intact ceramic layer, see Figure 8b. Furthermore, the blue failure appearance in the photograph suggests the presence of NiO and other spinels. The TGO layer thickness at failure, as shown in Figure 8d, was measured to be approximately 2 µm, which happens to be lower than the critical TGO thickness (as seen in TCF tested specimens).

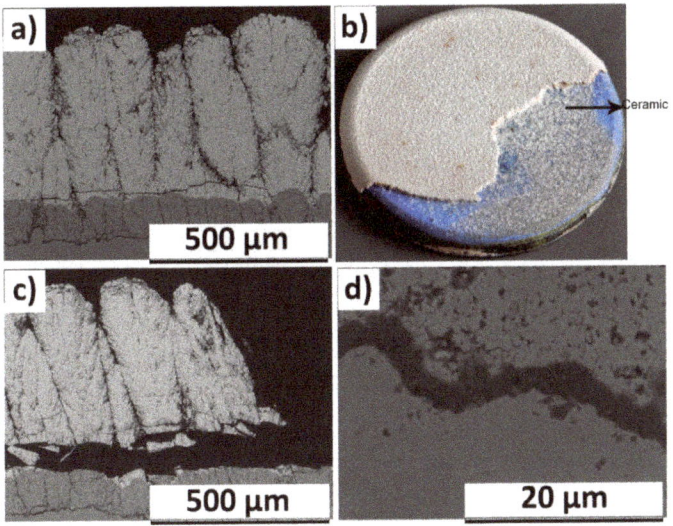

Figure 8. SEM micrograph of BRT (burner rig test) failed (814 cycles) 400GZ/100YSZ (**a**) Cross section; (**b**) photograph; (**c**) cross sectional view at a different location; (**d**) TGO.

In the case of failed 250GZ/250YSZ TBC, horizontal crack parallel to GZ/YSZ interface was observed. Furthermore, the location of horizontal crack shifted away (approximately 250 µm) from the bond coat compared to the 400GZ/100YSZ failed TBC, see Figures 8a and 9a. The reason for such a shift in failure location could be due to the presence of higher fracture toughness material, i.e., YSZ, at the previously reported failure location. However, the durability results did not show improvement over 400GZ/100YSZ. The photograph of failed 250GZ/250YSZ also confirmed the spallation in the ceramic layer, see Figure 9b. High magnification SEM micrograph of the GZ/YSZ interface showed interlinking of horizontal and vertical cracks in GZ layer, according to Figure 9c, which led to spallation of GZ layer. TGO thickness at failure in this case was measured to be less than 2 µm. Furthermore, the YSZ layer close to TGO was free from cracks, according to Figure 9d.

In the case of failed 100GZ/400YSZ TBC, the cross sectional SEM micrograph shows delamination of the top GZ layer along with a thin remnant layer of GZ, according to Figure 10a. The failure location shifted further away from the bond coat (approximately 400 µm from the bond coat). The BRT lifetime was determined based on the photographs captured after each cycle. In this case, the test specimen was exposed to BRT conditions for longer cycles than its lifetime. Therefore, after the spallation of the GZ layer, horizontal cracks in YSZ appeared at two different locations; one close to the free surface and one close to the bond coat, see Figure 10c,d. The reason for horizontal crack propagation close to the bond coat could be attributed to the mismatch in coefficient of thermal expansion (CTE) between YSZ and metallic substrate. When the stored elastic energy in the TBC due to CTE mismatch exceeds the fracture toughness of the material, a crack propagates through the coating, leading to spallation. Previous findings on failure of the YSZ single layer TBC when subjected to a burner rig test reported similar horizontal crack propagation in the YSZ layer close to the bond coat [14]. The photograph also suggests that the failure occurred in the ceramic layer, as some part of the ceramic was still intact after the test, according to Figure 10b.

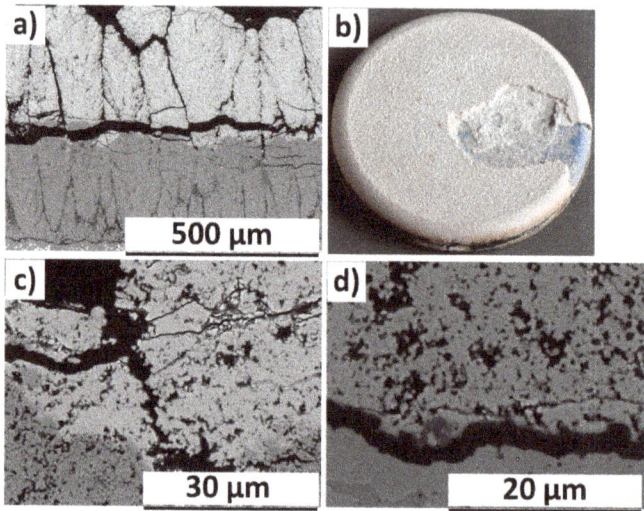

Figure 9. SEM micrograph of BRT failed (301 cycles) 250GZ/250YSZ (**a**) Cross section; (**b**) photograph; (**c**) high magnification cross sectional view; (**d**) TGO.

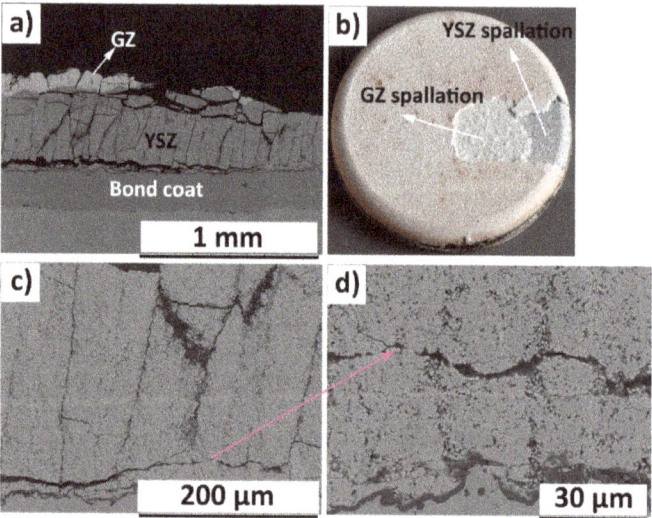

Figure 10. SEM micrograph of BRT failed (101 cycles) 100GZ/400YSZ (**a**) Cross section; (**b**) photograph; (**c**) cross sectional view at a different location; (**d**) TGO.

In the BRT, the relatively shorter time of exposure to high temperature during each cycle prevents sufficient time for oxidation of the bond coat and hence TGO growth is restricted, as seen in this work (<2 µm at failure in all the tested coatings). Simultaneously, the ceramic coating also undergoes sintering to some extent at a relatively higher exposure temperature, with the sintering rate being higher close to the surface than near the bond coat. The evidence of sintering was reported in our previous work for SPS-processed GZ and YSZ TBCs when subjected to BRT (same test rig) even at lower surface temperature (1300 °C) and time, see [14]. If the TBC failure occurs near the surface, it could be argued that the potential cause for failure was due to sintering. The failure analysis of BRT specimens in this study did not reveal failure at the TBC surface. Furthermore, the failure was not observed in the TGO (<2 µm thickness) layer under BRT conditions as the TGO thickness at failure was well below the critical TGO thickness On the other hand, the CTE mismatch between ceramic coating and the metallic substrate during BRT results in accumulation of strain energy in the TBC. When the strain energy in the coating exceeds the fracture toughness of the material, failure occurs. In the current work, all the three double-layered TBC variations failed in GZ (inferior fracture toughness material) layer close to interface with base YSZ layer, indicating CTE mismatch as the potential cause for failure. It seems that the three mechanisms (oxidation of bond coat, sintering and CTE mismatch) compete with each other for TBC failure during BRT and, in this case, the CTE mismatch between the ceramic and metallic substrate wins the race. Based on the author's hypothesis, with an increase in YSZ layer thickness in the GZ/YSZ double-layered TBC, the BRT lifetime should have been higher. However, the BRT results obtained in this work contradict the author's hypothesis, although the failure mechanisms were similar for all the three GZ/YSZ double-layered TBC variants. It is worthy to mention that GZ has lower thermal conductivity than YSZ (approximately 30% lower) [8,23]. Improved thermal insulation in 400GZ/100YSZ TBC due to lower YSZ thickness leads to lower bond coat temperature, which results in lower CTE mismatch and stress levels in 400GZ/100YSZ than 100GZ/400YSZ TBC. This could be one possible explanation for the improved durability of 400GZ/100YSZ than 100GZ/400YSZ. BRT results indicate that, in addition to the TBC fracture toughness, thermal insulation property of the TBC also plays an important role in governing durability.

4. Conclusions

In this work, gadolinium zirconate/YSZ double-layered TBCs with varying GZ/YSZ thickness combinations were investigated to evaluate the hypothesis that an increase in the YSZ layer thickness would enhance TBC durability. It was conjectured that this would provide a bigger region of higher toughness in the immediate vicinity of the bond coat to consequently delay cracking in the tougher YSZ layer or shift the probable failure-prone GZ region further away from the bond coat. The as-sprayed TBCs were subjected to different thermal cyclic test conditions, i.e., BRT (with temperature gradient) and TCF (without temperature gradient). Test results obtained in this work oppose the author's hypothesis, as an increase in YSZ layer thickness in the GZ/YSZ double-layered TBC led to inferior durability under BRT and TCF test conditions. A possible explanation for inferior durability of 100GZ/400YSZ under BRT could be due to its higher thermal conductivity than 400GZ/100GZ (GZ has 30% lower thermal conductivity than YSZ), resulting in severe stress state (due to higher CTE mismatch) in the 100GZ/400YSZ coating. On the other hand, in the case of TCF, inferior durability of TBC with higher YSZ thickness (100GZ/400YSZ) could be due to the higher oxygen penetration resistance of GZ than YSZ.

Failure modes under TCF and BRT conditions were found to differ in the investigated TBCs. Among the three possible mechanisms for TBC failure i.e., sintering of the ceramic, oxidation of bond coat and CTE mismatch between the top coat and bond coat; oxidation of the bond coat and reaching a critical TGO thickness were found to be the reasons for TBC failure under TCF test conditions. In contrast, in the case of BRT, it was shown that CTE mismatch between the ceramic coating and metallic substrate dictates the TBC failure. Furthermore, with an increase in the YSZ layer thickness in the GZ/YSZ double-layered TBC, failure location shifted northwards from the bond coat, but remained in the GZ layer close to the interface with YSZ. In this work, it was also shown that an increase in fracture toughness at the probable failure location does not necessarily improve the durability. Other factors (such as thermal conductivity of the top coat) could play an important role in dictating the durability of the TBC.

Although the failure modes under different thermal cyclic conditions differed for the investigated TBCs, durability was shown to be superior for 400GZ/100YSZ TBC. Further improvement in the durability of 400GZ/100YSZ TBC could be achieved by opting for a denser GZ microstructure (due to improved fracture toughness) as failure in BRT in this work was shown to be in the GZ layer close to the YSZ interface.

Author Contributions: Conceptualization by S.M., N.C. and N.M.; Spraying was done by S.B.; Writing-first draft by S.M.; Writing-review and editing by S.J., N.M. and N.C.

Funding: This research was funded by the knowledge foundation (KK stiftelsen), Grant number: Dnr-20140130, Sweden.

Acknowledgments: The authors would like to thank Robert Vassen for the fruitful discussions related to the burner rig test results. The authors are also grateful to Sigrid Schwartz and Martin Tandler from IEK-1, Forschungszentrum Jülich, Germany for performing the burner rig test. Financial support from SiCoMaP+ research school and KK foundation (Dnr: 20140130) is deeply acknowledged.

Conflicts of Interest: The authors declare no conflict of interest.

References

1. Stöver, D.; Funke, C. Directions of the development of thermal barrier coatings in energy applications. *J. Mater. Process. Techol.* **1999**, *92–93*, 195–202.
2. Ballard, J.D.; Davenport, J.; Lewis, C.; Doremus, R.H.; Schadler, L.S.; Nelson, W. Phase stability of thermal barrier coatings made from 8 wt. % yttria stabilized zirconia: A technical note. *J. Therm. Spray Technol.* **2003**, *12*, 34–37. [CrossRef]
3. Thompson, J.A.; Clyne, T.W. The effect of heat treatment on the stiffness of zirconia top coats in plasma-sprayed TBCs. *Acta Mater.* **2001**, *49*, 1565–1575. [CrossRef]
4. Aygun, A.; Vasiliev, A.L.; Padture, N.P.; Ma, X. Novel thermal barrier coatings that are resistant to high-temperature attack by glassy deposits. *Acta Mater.* **2007**, *55*, 6734–6745. [CrossRef]
5. Drexler, J.M.; Shinoda, K.; Ortiz, A.L.; Li, D.S.; Vasiliev, A.L.; Gledhill, A.D.; Sampath, S.; Padture, N.P. Air-plasma-sprayed thermal barrier coatings that are resistant to high-temperature attack by glassy deposits. *Acta Mater.* **2010**, *58*, 6835–6844. [CrossRef]
6. Gledhill, A.D.; Reddy, K.M.; Drexler, J.M.; Shinoda, K.; Sampath, S.; Padture, N.P. Mitigation of damage from molten fly ash to air-plasma-sprayed thermal barrier coatings. *Mater. Sci. Eng. A* **2011**, *528*, 7214–7221. [CrossRef]
7. Witz, G.; Shklover, V.; Steurer, W.; Bachegowda, S.; Bossmann, H.-P. High-temperature interaction of yttria stabilized zirconia coatings with CaO-MgO-Al$_2$O$_3$-SiO$_2$ (CMAS) deposits. *Surf. Coat. Technol.* **2015**, *265*, 244–249. [CrossRef]
8. Vaßen, R.; Jarligo, M.O.; Steinke, T.; Mack, D.E.; Stöver, D. Overview on advanced thermal barrier coatings. *Surf. Coat. Technol.* **2010**, *205*, 938–942. [CrossRef]
9. Krämer, S.; Yang, J.; Levi, C.G. Infiltration-inhibiting reaction of gadolinium zirconate thermal barrier coatings with CMAS melts. *J. Am. Ceram. Soc.* **2008**, *91*, 576–583. [CrossRef]
10. Mahade, S.; Zhou, D.; Curry, N.; Markocsan, N.; Nylén, P.; Vaßen, R. Tailored microstructures of gadolinium zirconate/YSZ multi-layered thermal barrier coatings produced by suspension plasma spray: Durability and erosion testing. *J. Mater. Process. Technol.* **2019**, *264*, 283–294. [CrossRef]
11. Leckie, R.M.; Krämer, S.; Rühle, M.; Levi, C.G. Thermochemical compatibility between alumina and ZrO$_2$-GdO$_{3/2}$ thermal barrier coatings. *Acta Mater.* **2005**, *53*, 3281–3292. [CrossRef]
12. Bakan, E.; Mack, D.E.; Mauer, G.; Vaßen, R. Gadolinium zirconate/YSZ thermal barrier coatings: Plasma spraying, microstructure, and thermal cycling behavior. *J. Am. Ceram. Soc.* **2014**, *97*, 4045–4051. [CrossRef]
13. Bakan, E.; Mack, D.E.; Mauer, G.; Mücke, R.; Vaßen, R. Porosity-Property Relationships of Plasma-Sprayed Gd$_2$Zr$_2$O$_7$/YSZ Thermal Barrier Coatings. *J. Am. Ceram. Soc.* **2015**, *98*, 2647–2654. [CrossRef]

14. Mahade, S.; Curry, N.; Björklund, S.; Markocsan, N.; Nylén, P.; Vaßen, R. Functional performance of Gd$_2$Zr$_2$O$_7$/YSZ multi-layered thermal barrier coatings deposited by suspension plasma spray. *Surf. Coat. Technol.* **2017**, *318*, 208–216. [CrossRef]
15. Zhong, X.; Zhao, H.; Zhou, X.; Liu, C.; Wang, L.; Shao, F.; Yang, K.; Tao, S.; Ding, C. Thermal shock behavior of toughened gadolinium zirconate/YSZ double-ceramic-layered thermal barrier coating. *J. Alloy. Compd.* **2014**, *593*, 50–55. [CrossRef]
16. Curry, N.; VanEvery, K.; Snyder, T.; Markocsan, N. Thermal conductivity analysis and lifetime testing of suspension plasma-sprayed thermal barrier coatings. *Coatings* **2014**, *4*, 630–650. [CrossRef]
17. Schulz, U.; Menzebach, M.; Leyens, C.; Yang, Y.Q. Influence of substrate material on oxidation behavior and cyclic lifetime of EB-PVD TBC systems. *Surf. Coat. Technol.* **2001**, *146–147*, 117–123. [CrossRef]
18. Zhang, P.; Yuan, K.; Peng, R.L.; Li, X.-H.; Johansson, S. Long-term oxidation of MCrAlY coatings at 1000 °C and an Al-activity based coating life criterion. *Surf. Coat. Technol.* **2017**, *332*, 12–21. [CrossRef]
19. Mahade, S.; Curry, N.; Jonnalagadda, K.P.; Peng, R.L.; Markocsan, N.; Nylén, P. Influence of YSZ layer thickness on the durability of gadolinium zirconate/YSZ double-layered thermal barrier coatings produced by suspension plasma spray. *Surf. Coat. Technol.* **2019**, *357*, 456–465. [CrossRef]
20. Wellman, R.G.; Nicholls, J.R. A review of the erosion of thermal barrier coatings. *J. Phys. D Appl. Phys.* **2007**, *40*, R293–R305. [CrossRef]
21. Mahade, S.; Ruelle, C.; Curry, N.; Holmberg, J.; Björklund, S.; Markocsan, N.; Nylén, P. Understanding the effect of material composition and microstructural design on the erosion behavior of plasma sprayed thermal barrier coatings. *Appl. Surf. Sci.* **2019**, *488*, 170–184. [CrossRef]
22. ImageJ. Available online: https://imagej.en.softonic.com (accessed on 8 April 2019).
23. Mahade, S.; Curry, N.; Björklund, S.; Markocsan, N.; Nylén, P. Thermal conductivity and thermal cyclic fatigue of multilayered Gd$_2$Zr$_2$O$_7$/YSZ thermal barrier coatings processed by suspension plasma spray. *Surf. Coat. Technol.* **2015**, *283*, 329–336. [CrossRef]
24. Traeger, F.; Vaßen, R.; Rauwald, K.-H.; Stöver, D. Thermal cycling setup for testing thermal barrier coatings. *Adv. Eng. Mater.* **2003**, *5*, 429–432. [CrossRef]
25. Mahade, S.; Curry, N.; Björklund, S.; Markocsan, N.; Nylén, P. Failure analysis of Gd$_2$Zr$_2$O$_7$/YSZ multi-layered thermal barrier coatings subjected to thermal cyclic fatigue. *J. Alloy. Compd.* **2016**, *689*, 1011–1019. [CrossRef]
26. Cheng, B.; Yang, N.; Zhang, Q.; Zhang, M.; Zhang, Y.-M.; Chen, L.; Yang, G.J.; Li, C.X.; Li, C.J. Sintering induced the failure behavior of dense vertically crack and lamellar structured TBCs with equivalent thermal insulation performance. *Ceram. Int.* **2017**, *43*, 15459–15465. [CrossRef]
27. Cheng, B.; Zhang, Y.M.; Yang, N.; Zhang, M.; Chen, L.; Yang, G.J.; Li, C.X.; Li, G. Sintering-induced delamination of thermal barrier coatings by gradient thermal cyclic test. *J. Am. Ceram. Soc.* **2017**, *100*, 1820–1830. [CrossRef]
28. Cao, X.Q.; Vassen, R.; Stoever, D. Ceramic materials for thermal barrier coatings. *J. Eur. Ceram. Soc.* **2004**, *24*, 1–10. [CrossRef]
29. Mahade, S.; Li, R.; Curry, N.; Björklund, S.; Markocsan, N.; Nylén, P. Isothermal oxidation behavior of Gd$_2$Zr$_2$O$_7$/YSZ multilayered thermal barrier coatings. *Int. J. Appl. Ceram. Technol.* **2016**, *13*, 443–450. [CrossRef]
30. Dong, H.; Yang, G.-J.; Li, C.-X.; Luo, X.-T.; Li, C.-J. Effect of TGO thickness on thermal cyclic lifetime and failure mode of plasma-sprayed TBCs. *J. Am. Ceram. Soc.* **2014**, *97*, 1226–1232. [CrossRef]
31. Lu, Z.; Myoung, S.-W.; Jung, Y.-G.; Balakrishnan, G.; Lee, J.; Paik, U. Thermal fatigue behavior of air-plasma sprayed thermal barrier coating with bond coat species in cyclic thermal exposure. *Mater. Basel Switz.* **2013**, *6*, 3387–3403. [CrossRef]
32. Smialek, J.L. Compiled furnace cyclic lives of EB-PVD thermal barrier coatings. *Surf. Coat. Technol.* **2015**, *276*, 31–38. [CrossRef]
33. Gupta, M.; Markocsan, N.; Rocchio-Heller, R.; Liu, J.; Li, X.-H.; Östergren, L. Failure analysis of multilayered suspension plasma-sprayed thermal barrier coatings for gas turbine applications. *J. Therm. Spray Technol.* **2018**, *27*, 402–411. [CrossRef]
34. Karunaratne, M.S.A.; Kyaw, S.; Jones, A.; Morrell, R.; Thomson, R.C. Modelling the coefficient of thermal expansion in Ni-based superalloys and bond coatings. *J. Mater. Sci.* **2016**, *51*, 4213–4226. [CrossRef]

35. Viswanathan, V.; Dwivedi, G.; Sampath, S. Multilayer, multimaterial thermal barrier coating systems: Design, synthesis, and performance assessment. *J. Am. Ceram. Soc.* **2015**, *98*, 1769–1777. [CrossRef]
36. Zhou, D.; Guillon, O.; Vaßen, R. Development of YSZ thermal barrier coatings using axial suspension plasma spraying. *Coatings* **2017**, *7*, 120. [CrossRef]

© 2019 by the authors. Licensee MDPI, Basel, Switzerland. This article is an open access article distributed under the terms and conditions of the Creative Commons Attribution (CC BY) license (http://creativecommons.org/licenses/by/4.0/).

Article

A Study on the Microstructural Characterization and Phase Compositions of Thermally Sprayed Al_2O_3-TiO_2 Coatings Obtained from Powders and Water-Based Suspensions

Monika Michalak [1], Filofteia-Laura Toma [2], Leszek Latka [1], Pawel Sokolowski [1,*], Maria Barbosa [2] and Andrzej Ambroziak [1]

[1] Faculty of Mechanical Engineering, Wroclaw University of Science and Technology, 50-371 Wroclaw, Poland; monika.michalak@pwr.edu.pl (M.M.); leszek.latka@pwr.edu.pl (L.L.); andrzej.ambroziak@pwr.edu.pl (A.A.)
[2] Fraunhofer-Institute for Material and Beam Technology (IWS) Dresden, 01277 Dresden, Germany; Filofteia-Laura.Toma@iws.fraunhofer.de (F.-L.T.); maria.barbosa@iws.fraunhofer.de (M.B.)
* Correspondence: pawel.sokolowski@pwr.edu.pl

Received: 27 April 2020; Accepted: 4 June 2020; Published: 9 June 2020

Abstract: In this work, the alumina (Al_2O_3) and alumina-titania coatings with different contents of TiO_2, i.e., Al_2O_3 + 13 wt.% TiO_2 and Al_2O_3 + 40 wt.% TiO_2, were studied. The coatings were produced by means of powder and liquid feedstock thermal spray processes, namely atmospheric plasma spraying (APS), suspension plasma spraying (SPS) and suspension high-velocity oxygen fuel spraying (S-HVOF). The aim of the study was to investigate the influence of spray feedstocks characteristics and spray processes on the coating morphology, microstructure and phase composition. The results revealed that the microstructural features were clearly related both to the spray processes and chemical composition of feedstocks. In terms of phase composition, in Al_2O_3 (AT0) and Al_2O_3 + 13 wt.% TiO_2 (AT13) coatings, the decrease in α-Al_2O_3, which partially transformed into γ-Al_2O_3, was the dominant change. The increased content of TiO_2 to 40 wt.% (AT40) involved also an increase in phases related to the binary system Al_2O_3-TiO_2 (Al_2TiO_5 and $Al_{2-x}Ti_{1+x}O_5$). The obtained results confirmed that desired α-Al_2O_3 or α-Al_2O_3, together with rutile-TiO_2 phases, may be preserved more easily in alumina-titania coatings sprayed by liquid feedstocks.

Keywords: Al_2O_3-TiO_2 system; APS; suspension spraying; microstructure; morphology; phase composition

1. Introduction

Thermal spraying is a well-known technique used for the deposition of different types of coatings for applications in many industrial fields. The processes based on direct spraying of liquid feedstocks have gained increasing interest in recent years. Among different technologies, the most intensively studied are suspension plasma spraying (SPS) and suspension high-velocity oxygen fuel spraying (S-HVOF) [1,2], patented, respectively, by Gitzhofer et al. in 1997 [3] and Gadow et al. in 2011 [4]. Since that time, these techniques have been developed in parallel [5,6]. By spraying with feedstocks of submicrometer- or even nanometer-sized powders, the microstructural features of coatings change significantly and, thus, the different functional properties of coatings may be improved [7,8].

Al_2O_3 is among the most popular oxide ceramic materials used in thermal spray technology. Besides pure Al_2O_3, the attention is paid nowadays to the Al_2O_3 + TiO_2 ceramics, including especially Al_2O_3 + 13 wt.% TiO_2 (due to its outstanding tribological behavior) and Al_2O_3 + 40 wt.% TiO_2 (e.g., for its improved fracture toughness). In general, with the increased content of TiO_2, the melting ability of the powders is more favorable and, then, the deposition of denser and defect-free coatings is easier.

Furthermore, the addition of titania to alumina coatings improves, e.g., fracture toughness, and this may be used to improve the wear resistance of alumina-based coatings. The beneficial increase in TiO_2 was confirmed, e.g., at a content of 44 wt.% TiO_2 where the formation of the Al_2TiO_5 phase takes place, which is of better corrosion resistance in dilute acids [9].

Furthermore, when compared to conventional powder thermal spray processes, suspension-based techniques show higher flexibility in tailoring the microstructure and chemical composition of the feedstock material [2]. For example, APS-sprayed coatings based on Al_2O_3 and Al_2O_3 + TiO_2 are relatively porous which is undesirable in some applications, for example in electronics and sealing systems [10,11]. Suspension-based coatings seem to be an interesting solution for such problems. The studies devoted to the liquid feedstock spraying [12–14] have already shown that these processes can provide thinner coatings with comparable, or even better properties, including hardness, corrosion or wear resistance.

Studies on Al_2O_3 sprayed by S-HVOF [15,16] have shown that these coatings are of higher density and improved adhesion, and contain refined microstructure with small lamellas when compared to conventional APS or HVOF [17]. Another benefit of suspension spraying using HVOF is the possibility of retention of the original crystalline phase. This is of great importance especially in the case of Al_2O_3 because, during spraying, this material tends to transform from initial thermodynamically stable α-phase into metastable γ-phase, which is characterized, e.g., by lower corrosion resistance [18].

It should be emphasized that most of the articles devoted to the spraying from liquid feedstocks concern SPS [19,20] and S-HVOF [21,22] of Al_2O_3 coatings. There are only a few papers that consider the influence of TiO_2 addition in such coatings. Darut et al. [23] investigated phase transformation in Al_2O_3 + 13 wt.% TiO_2 SPS coatings, while Vicent et al. [24] characterized microstructure and nanoindentation properties, also in similar coatings.

In the article, both powder- and suspension-based feedstocks of Al_2O_3 (AT0), Al_2O_3 + 13 wt.% TiO_2 (AT13) and Al_2O_3 + 40 wt.% TiO_2 (AT40) were sprayed by means of: (i) APS, (ii) SPS and (iii) S-HVOF. It is well-known that feedstock characteristics have a great influence on the splat formation during spraying [8]. Depending on the size of raw powder, manufacturing method of the suspensions, particle size distribution, etc., different microstructural properties may be obtained [15]. Therefore, all suspensions were formulated under laboratory conditions in a repetitive manner. The obtained coatings were analyzed and compared in the terms of morphology, microstructure and phase composition.

2. Materials and Experimental Methods

2.1. Feedstocks

In this study, the powders with 3 different chemical compositions were used: Al_2O_3, Al_2O_3 + 13 wt.% TiO_2 and Al_2O_3 + 40 wt.% TiO_2. They are labeled and denoted within the article as AT0, AT13 and AT40, respectively.

The commercially available AT0, AT13 and AT40 spray powders manufactured by Oerlikon Metco (Pfäffikon, Switzerland) were used to produce coatings by means of conventional APS: (i) Al_2O_3 Metco 6103, in agglomerated and sintered form, with the particle size −45 + 15 µm; (ii) Al_2O_3-13TiO_2 Metco 6221, in agglomerated and sintered form, with the particle size −45 + 15 µm; and (iii) Al_2O_3-40TiO_2 Metco 131VF, in agglomerated form, with the particle size of −45 + 5 µm. The powder particle size distribution was verified by the means of powder granulometry, by Partica LA-950V2 (Horiba, Kyoto, Japan), according to the standard [25].

For the formulation of liquid feedstocks, the Al_2O_3-TiO_2 powders were milled using a high-energy ball milling EMax setup (Retsch GmbH, Haan, Germany) for 80 min per batch. However, the Al_2O_3 suspension was formulated by using commercially available α-Al_2O_3 submicrometer-sized powder MARTOXID® MZS-1 (Martinswerk GmbH, Bergheim, Germany), labeled below as AT0*. The use of commercial AT0* powder was caused by the difficulties in formulating stable suspension based on milled Metco 6103 (AT0) powder. All suspensions used within this study were water-based and contained 25 wt.% of submicrometer-sized solids.

The detailed results of powder granulometry measurements are listed in Table 1. The morphology of powders was investigated with the use of SEM microscope JEOL JSM-6610A (JEOL, Tokyo, Japan).

Table 1. Particle size distribution of Al_2O_3 raw powder and milled Al_2O_3-TiO_2 powders for suspension preparation; d_v—particle size by volume [μm].

	AT0* SPS/S-HVOF	AT13 SPS/S-HVOF	AT40 SPS/S-HVOF
d_{v10}	0.81 μm	0.67 μm	0.51 μm
d_{v50}	1.22 μm	1.15 μm	0.67 μm
d_{v90}	1.82 μm	1.73 μm	1.01 μm

The formulated suspensions were further investigated mainly in terms of rheological properties. The measurements were carried out by modular compact rheometer MCR 72 (Anton Paar, Graz, Austria) in cone-plate (CP) rotation mode, in order to estimate the viscosity and shear stress. Values of pH were measured using HI-2002 Edge pH Meter (Hanna Instruments, Leighton Buzzard, UK).

2.2. Deposition Process

The 304 austenitic stainless steel coupons (25 mm diameter, 2 mm thick) were used as substrates. Just before spraying, the substrates were sand-blasted with corundum and cleaned with ethanol. The Ni20Cr (Amperit 250, Höganäs Germany GmbH, Laufenburg, Germany) bond coats of thickness about 70 μm were previously deposited by APS. Then, the alumina and alumina-titania topcoats were fabricated by means of APS (spraying was performed at Wrocław University of Science and Technology, Poland), SPS, and S-HVOF (both suspension spraying trials were done at Fraunhofer IWS, Dresden, Germany). All liquid feedstocks were fed using the industrially suitable suspension feeder. The feeder was equipped with continuous suspension stirring and controlled pressure/suspension flow rates. It was developed by Fraunhofer IWS and tested already with a wide variety of suspensions [6,26]. Prior to the spraying, the suspensions were continuously mechanically stirred in order to redisperse feedstocks and to avoid any clogging in the suspension lines.

2.2.1. Atmospheric Plasma Spraying (APS)

Conventional atmospheric plasma spraying was carried out using one cathode, one anode SG-100 gun (Praxair, IN, USA). The spraying of each powder was preceded by the optimization of the deposition parameters. The details can be found elsewhere [27,28]. The spraying parameters, considered in the presented study, are given in Table 2.

Table 2. APS spraying parameters.

Spray Variables	AT0	AT13	AT40
Electrical power, kW		35	
Ar/H_2, L·min^{-1}		45/5	
Spray distance, mm		100	
Relative torch scan velocity, m·s^{-1}		0.3	
Powder feed rate, g·min^{-1}		20	
Coating thickness, μm		200–250	
Thickness per pass, μm/pass		29–35	

Prior to the spraying, powders were dried in the temperature of 120 °C within 2 h, in order to avoid clogging in the powder liner or injector. Powders were injected radially, with the external feedstock injection mode.

2.2.2. Suspension Plasma Spraying (SPS)

Suspension plasma spraying was carried out with the use of cascade KK plasma gun (AMT AG Kleindöttingen, Switzerland) with a 7 mm nozzle and Ar/H$_2$ plasma gas mixture. It should be noticed that this configuration allowed using relatively long spray distance, which was very similar as in the case of APS or S-HVOF. In the SPS process, the suspensions were externally and radially injected. All process parameters are collected in Table 3.

Table 3. SPS spraying parameters.

Spray Variables	AT0*	AT13	AT40
Electrical power, kW		70	
Ar/H$_2$, L·min^{-1}		50/6	
Spray distance, mm		80	
Relative torch scan velocity, m·s^{-1}		0.8	
Suspension feed rate, mL·min^{-1}	35	35	42
Coating thickness, µm		200–250	
Thickness per pass, µm/pass		9–13	

2.2.3. Suspension High-Velocity Oxygen Fuel Spraying (S-HVOF)

The S-HVOF process was performed by using the Top Gun setup (GTV Verschleißschutz GmbH, Luckenbach, Germany). The combustion chamber of conventional HVOF Top Gun torch was modified, so the suspensions were injected internally and axially. The 8 mm diameter and 135 mm length nozzle were used each time with an ethylene/oxygen working gas mixture. The main spraying parameters are given in Table 4.

Table 4. Suspension high-velocity oxygen fuel spraying (S-HVOF) spraying parameters.

Spray Variables	AT0*	AT13	AT40
C$_2$H$_4$/O$_2$, L·min^{-1}	75/230	75/230	65/200
Spray distance, mm		90	
Relative torch scan velocity, m·s^{-1}		1.6	
Suspension feed rate, mL·min^{-1}		35	
Coating thickness, µm		200	
Thickness per pass, µm/pass		10–12	

2.3. Sample Characterization

The coatings' surfaces and cross-sections were investigated by scanning electron microscope SEM Phenom G2 Pro (Phenom World BV, Eindhoven, The Netherlands). In order to estimate the porosity of the coatings, the micrographs were analyzed by ImageJ software, according to the standard ASTM E2109-01 [29]. Porosity was estimated on the images taken at 1000× magnification and the average porosity was calculated based on at least 20 micrographs. The thickness of the coatings was analyzed on the micrographs taken at 500× magnification. At least 5 measurements in random regions were made for that purpose.

Phase compositions of the feedstock powders and coatings were determined by the X-ray diffraction technique (XRD) using the Empyrean diffractometer (Malvern Panalytical, Egham, UK) and with CuKα radiation. The measurements were performed in the range of 2θ equal to 10–80°, with 0.1° step size and 0.9 s/step counting time. The crystalline phases were identified using the JCPDS standard cards: 00–046–1212 (α-Al$_2$O$_3$), 00–010–0425 (γ-Al$_2$O$_3$), 00–041–0258 (Al$_2$TiO$_5$) and 00–21–1276 (rutile-TiO$_2$). The percentage of phases was determined by the method called reference intensity ratio (RIR), described in [30,31]. As for example, the contents of α-Al$_2$O$_3$ and γ-Al$_2$O$_3$ for AT0 coatings were determined with the use of Equation (1) [32]:

$$C_\gamma^{Al_2O_3} = \frac{I_\gamma(400)}{I_\alpha(113) + I_\gamma(400)} \cdot 100 \ [\%] \qquad (1)$$

where $C_\gamma^{Al_2O_3}$ is the γ-Al$_2$O$_3$ phase content, I_{hkl} is the intensity of the peak diffraction for the corresponding plane of a given phase.

3. Results and Discussion

3.1. Feedstocks

The APS spray powders (Figure 1a–c) were of micrometer sizes and spherical shape. AT0 and AT13 powders (Figure 1d), showed slightly greater particles when compared to AT40 powder. Furthermore, the microscopic investigations of feedstocks showed also that some AT40 particles were already fragmented in the delivery state. This phenomenon was not observed for AT0 and AT13 powders because those materials were not only agglomerated but sintered as well. This showed that the sintering of powders provides increased cohesion and thus, better flowability of powders during spraying. Indeed, during APS trials, the AT0 and AT13 powders showed good flowability and coatings were easily deposited. On the contrary, when spraying AT40 powder, deposition trials had to be repeated due to clogging of the transportation lines or injectors. Finally, all APS coatings were successfully deposited.

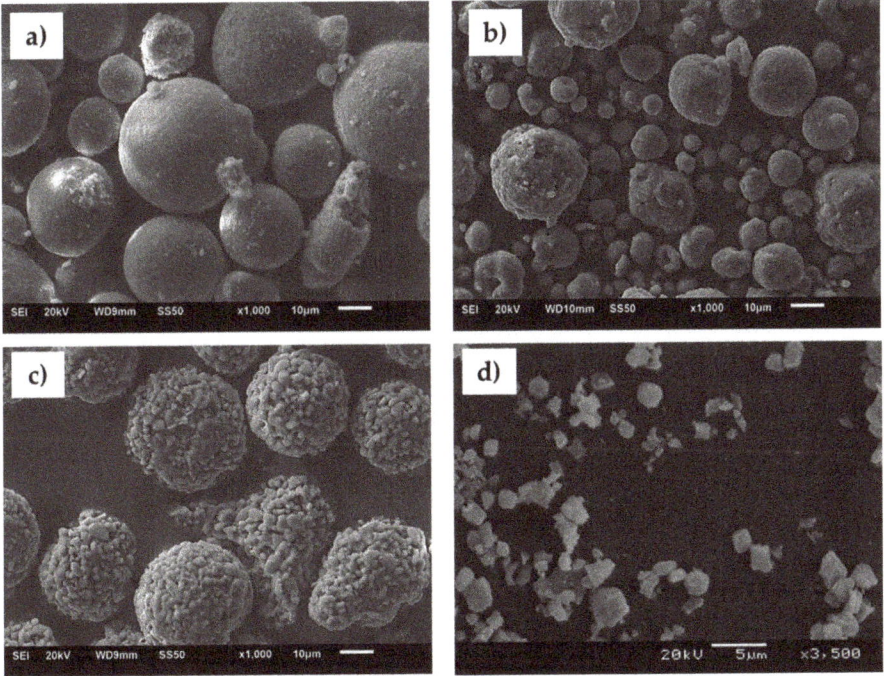

Figure 1. Exemplary morphology of powders used for spraying: (**a**) AT0 spray powder, (**b**) AT13 spray powder, (**c**) AT40 spray powder, (**d**) milled AT13 powder for suspension spraying.

All powders used for the suspension formulation were of d_{v50} of around 1 μm. The powders were very similar in terms of morphology; they revealed irregular crushed form and monomodal particle size so the micrograph of the representative AT13 powder was presented only (Figure 1d).

Figure 2a presents the relationship between the viscosity and shear rate of all home-made suspensions. The measured values were below 10 mPa·s at the shear rate of 100 s^{-1}, which was reported to be appropriate for a constant and stable feeding [6]. At higher shear rates the slight increase in viscosity was observed. However, the viscosity values were found to be still in a proper range [33].

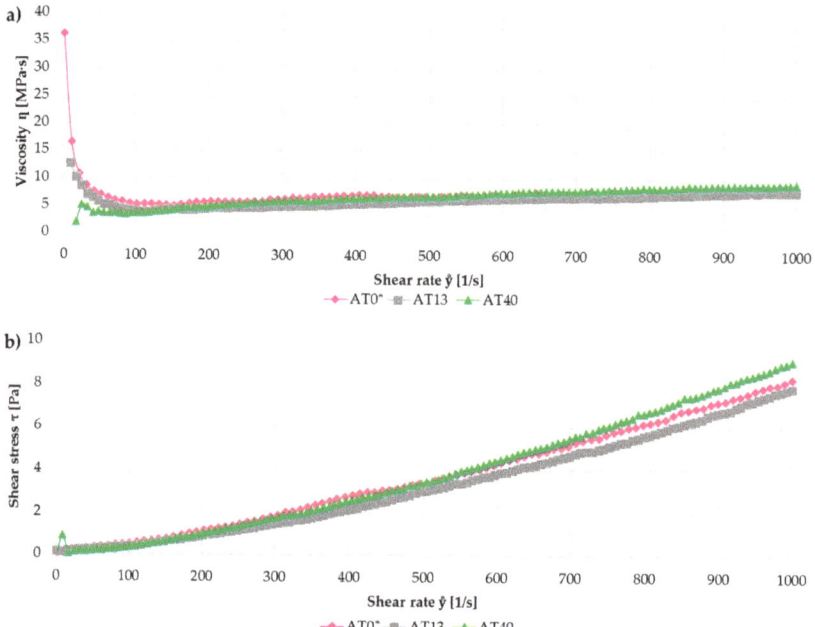

Figure 2. Comparison of viscosity (a) and shear stress (b) of AT0*, milled AT13 and AT40 water-based suspensions.

Another important factor to be considered was the value of pH. It should be ideally between 4 and 10, in order to prevent the hardware parts against the corrosion [6]. The measured values of all prepared suspensions were within this range (4 to 9.5). Finally, during SPS and S-HVOF spraying, the use of integrated stirrers in the pressurized vessels during suspension feeding limited the sedimentation of the feedstock and provided its continuous supply.

3.2. Morphology and Microstructure of the Coatings

The microstructural observations of coatings morphology (Figures 3 and 4) revealed clear differences between APS, SPS and S-HVOF deposits.

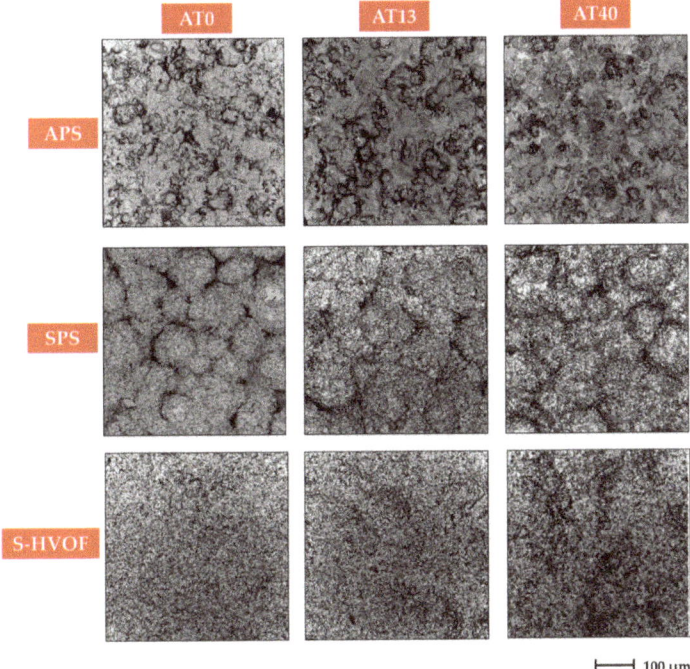

Figure 3. Low-magnification SEM images of coatings surface morphology, mag. 500×.

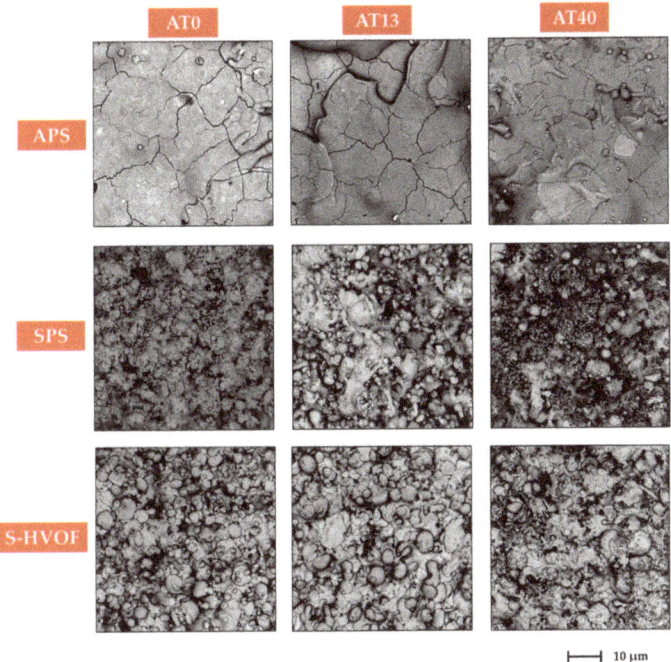

Figure 4. High-magnification SEM images of coatings surface morphology, mag. 5000×.

Coatings produced by APS showed the classical microstructure of the conventional thermally sprayed coatings; the presence of semimelted (or even nonmelted) powder particles and large, micron-sized splats with cracks (see Figures 3 and 4) could be observed.

SPS coatings revealed finely-grained morphology. However, in AT0 SPS a cauliflower-like topography was clearly noticed. The surfaces of AT13 and AT40 coatings were characterized by a more compact structure, but still relatively developed (Figure 5). Moreover, in SPS coatings, between melted lamellas, fine particles of sintered or partially melted powders were observed (Figure 4).

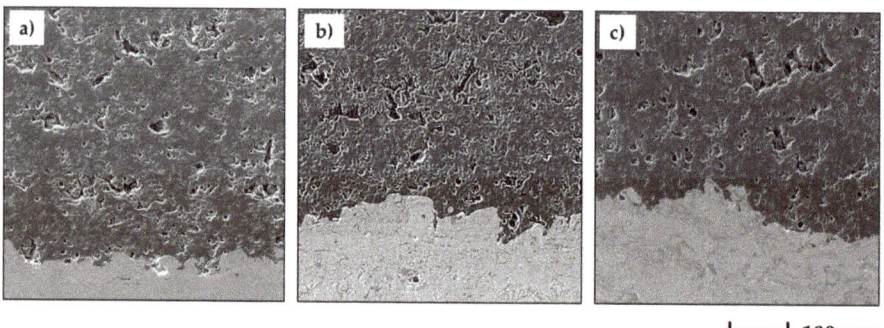

Figure 5. SEM images of cross-sections of APS-sprayed coatings: AT0 (**a**), AT13 (**b**), AT40 (**c**); mag. 1000×.

The results obtained for SPS coatings showed that further optimization of spraying parameters is needed, especially for AT0 SPS coating, due to its heterogeneous topography. Seshadri et al. [34], who characterized the conventional and cascaded arc plasma sprayed coatings, had shown that there is some threshold of, e.g., powder feed rate, where particles are not well melted or impinge the substrate in a partially melted state. If the optimum rate is highly exceeded, it could be expected that coatings may not be homogeneous in terms of microstructure and the porosity of coating may increase.

Additionally, the abovementioned morphology may be related to the type of solvent used for suspension formulation. Water requires approximately 3.2 times more energy for vaporization than ethyl alcohol [35]. On the other hand, water-based solvents are preferred due to their safe storage and handling as well as economic and environmental purposes [33,36]. Moreover, Vicent et al. [24] showed that power requirements become lower in the case of water-based suspension when solid content is higher (but it results, at the same time, in increased viscosity, which should be also adapted for the specified torch). It clearly shows that the optimization of spraying parameters is a complex task.

Surface micrographs showed that S-HVOF coatings were of most finely-grained microstructure, when compared with both APS and SPS. As expected, for this deposition method, the obtained coatings were characterized by a homogeneous and denser structure.

It should be noted that both KK and Top Gun torches enabled the deposition of suspension-based coatings, with the use of similar stand-off distances when compared to the SG-100 torch. For APS spraying, the distance was 100 mm, while for SPS and S-HVOF—80 mm and 90 mm, accordingly. Moreover, suspension-based processes had comparable deposition efficiencies as APS. Powder feed rate in atmospheric plasma spraying was about 20 g/min, while in SPS and S-HVOF spraying was equal to 25–35 mL/min. It provided feeding of the solid in the range of 10–15 g/min. When considering also the particle size (1 μm in suspension spraying, 20–30 μm in APS), it may be concluded that SPS and S-HVOF spraying had even better deposition efficiency as APS.

The cross-section micrographs (Figures 5 and 6) showed clear differences between APS and suspension sprayed coatings.

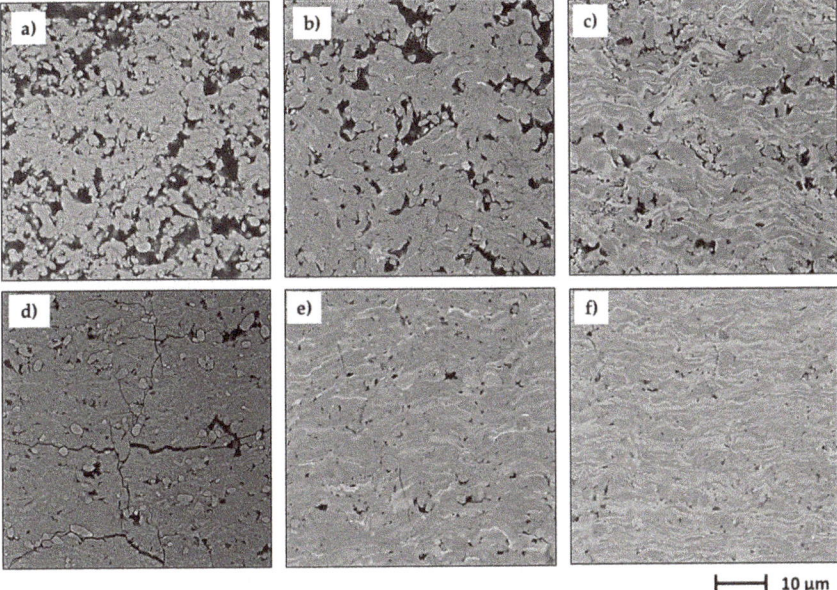

Figure 6. SEM images of cross-sections of suspension sprayed coatings: AT0 SPS (**a**), AT13 SPS (**b**), AT40 SPS (**c**), AT0 S-HVOF (**d**), AT13 S-HVOF (**e**), AT40 S-HVOF (**f**); mag. 5000×.

It was observed that APS coatings had pores of the greatest size, in the micrometer range. However, the mean volume area of pores, at a magnification of 1000× [29], was the highest for SPS coatings, as shown in Figure 5 (19 vol.% for AT0 SPS, 14 vol.% for AT13 SPS, 7 vol.% for AT40 SPS). Pores observed in those coatings were one order of magnitude smaller than in APS coatings. The densest and homogeneous coatings were observed for the S-HVOF process (1 vol.% for AT0 S-HVOF and AT40 S-HVOF, 3 vol.% for AT40 S-HVOF). However, in all types of S-HVOF coatings, both vertical and horizontal cracks were observed (Figure 6). In the case of AT0 S-HVOF coating, cracks constituted even more area (3 vol.%) than pores (1 vol.%). Regardless of the deposition method, it was observed that the addition of TiO$_2$ resulted in decreased coatings' porosity, see Figure 7. It was consistent with other results given in the literature [37,38].

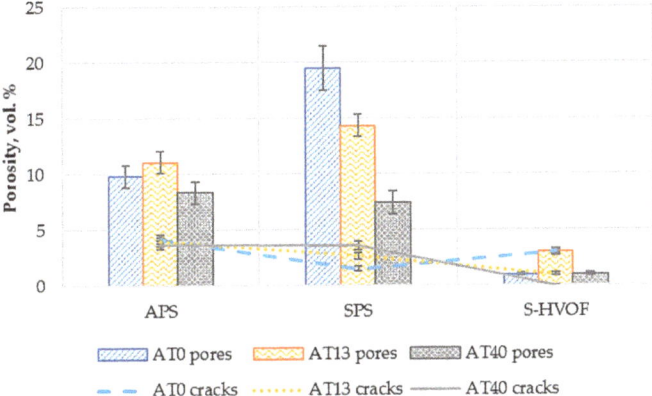

Figure 7. Average porosity in obtained coatings.

The low-magnification images (Figure 5) revealed that all types of coatings were well bonded to the bond coats. The micrographs taken at higher magnification showed that in some coatings, the fractions of nonmelted powders were present (Figure 6). They were found mainly in APS and SPS coatings; nevertheless, AT0 S-HVOF coating also revealed the presence of fine particles.

Fine fractions of powders, observed mainly in the cross-section images of SPS coatings (Figure 6), confirmed the presence of nonmelted or partially melted particles of the size around 1 µm, identified also in the top view images (Figures 3 and 4). It was due to the fact that powder did not penetrate into the plasma hot core and then was not sufficiently heated. The smaller powders have lower momentum and do not penetrate plasma jets in the same manner as bigger powders [39].

A close examination of the micrographs showed very different characteristics of the coatings. The microstructure of S-HVOF coatings, characterized by dense and uniformly distributed splats, was the result of high kinetic energy, typical for HVOF. In turn, more rough morphology and higher porosity, obtained in coatings sprayed by APS and SPS were influenced by plasma fluctuation, which is not to be neglected especially for SPS. However, in a cascade plasma gun, the electric arc was more stabilized than in classical plasma guns, like SG-100.

Different scales porosity was observed in the coatings. The detailed study on the porosity was already presented in our previous work [40]. According to the results, submicrometer- and micrometer-sized porosity was the highest for the coatings sprayed by SPS and the lowest for the S-HVOF deposits. The obtained porosity values were of similar range as reported in the literature [22,41]. The significant decrease in the porosity of AT40 coatings was observed for each spraying technique. This was mostly caused by the lower melting temperature of Al_2O_3 + 40 wt.% TiO_2 powders, when compared to pure Al_2O_3 or Al_2O_3 + 13 wt.% TiO_2. Moreover, agglomerated and nonsintered state of AT40 powders, also favored melting of this material; easily fragmented particles, due to their decreased diameter and mass, could be fast and well melted. It was also relevant for the formation of dense AT40 coatings, sprayed by S-HVOF.

3.3. Phase Composition

3.3.1. Micrometer- and Submicrometer-Sized Powders

Phase compositions of micrometer-sized powders in the delivery condition are shown in Figure 8. According to the results, AT0 powder consisted of a 100% stable α-Al_2O_3 phase. AT13 and AT40 powders, beyond α-Al_2O_3, contained also peaks of rutile-TiO_2 (AT13, AT40), tialite Al_2TiO_5 (AT13, AT40) and Al-rich solid solution $Al_{2-x}Ti_{1+x}O_5$ (AT40). Similar phase compositions of such powders were stated in the works of other authors [42,43].

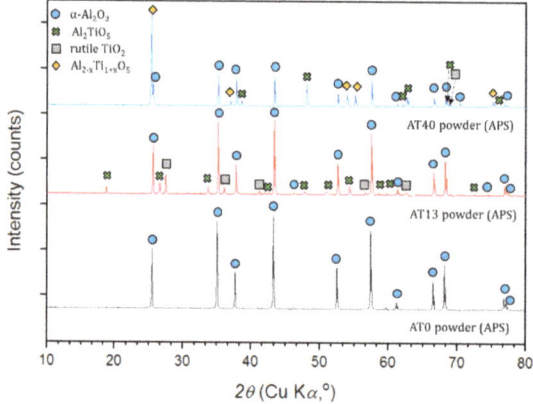

Figure 8. XRD patterns of powders used for APS spraying.

Figure 9 presents phases identified in AT0, AT13 and AT40 powders, dedicated to suspension preparation. As planned, the initial phase composition of (i) AT0 APS powder and (ii) AT0 SPS and S-HVOF powder was identical (100% α-Al$_2$O$_3$). This phase composition was also confirmed by other authors working with similar powders [44].

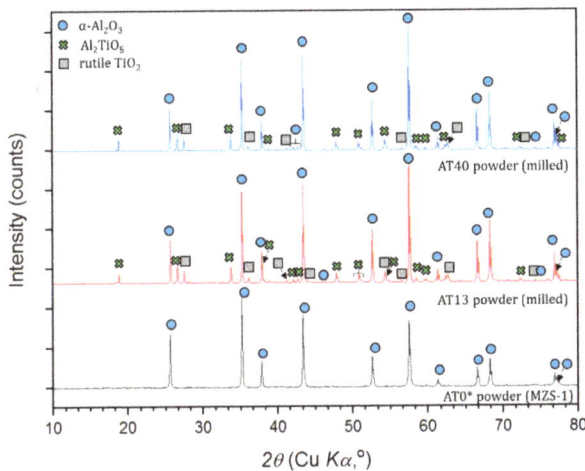

Figure 9. XRD patterns of raw AT0* and milled AT13, AT40 powders used for suspension spraying.

Special attention was paid to the phase composition of powders subjected to high-energy ball milling (AT13 and AT40). It is known already that such processes, including, e.g., high plastic deformation of powders may result in the phase transformation [45,46]. In the case of AT13 powders, no significant differences in the phase content were observed. In both cases, the identified phases were: α-Al$_2$O$_3$, Al$_2$TiO$_5$ and rutile-TiO$_2$. The content of phases was also quite similar in both powders. On the other hand, the XRD analysis showed that AT40 powder underwent a phase transformation during preprocessing. High-energy ball milling induced the Al$_{2-x}$Ti$_{1+x}$O$_5$ intermediate phase decomposition, at the expense of increased content of α-Al$_2$O$_3$, and, importantly, the formation of rutile-TiO$_2$. Bégin-Colin et al. [46], who studied the process of high-energy ball milling of TiO$_2$ powders, showed that phase transformations during this process are dependent, e.g., on grinding time. According to the results [46,47], the content of rutile-TiO$_2$ increases with the milling time and is additionally accompanied by the formation of high-pressure TiO$_2$(II). Its intensity increases first and then decreases with milling time. After about 70 min of milling, TiO$_2$(II) fully transforms, which induces a continuous increase in rutile-TiO$_2$ content. The results correspond well with the abovementioned studies—in this case, after 80 min of milling, rutile-TiO$_2$ peaks were well identified in AT40 submicrometer-sized powder.

APS, SPS and S-HVOF spraying resulted in the change of the coatings' phase composition, which was dependent both on (i) the chemical composition, as well as on (ii) the spraying technique. The quantitative results are summarized in Figure 10.

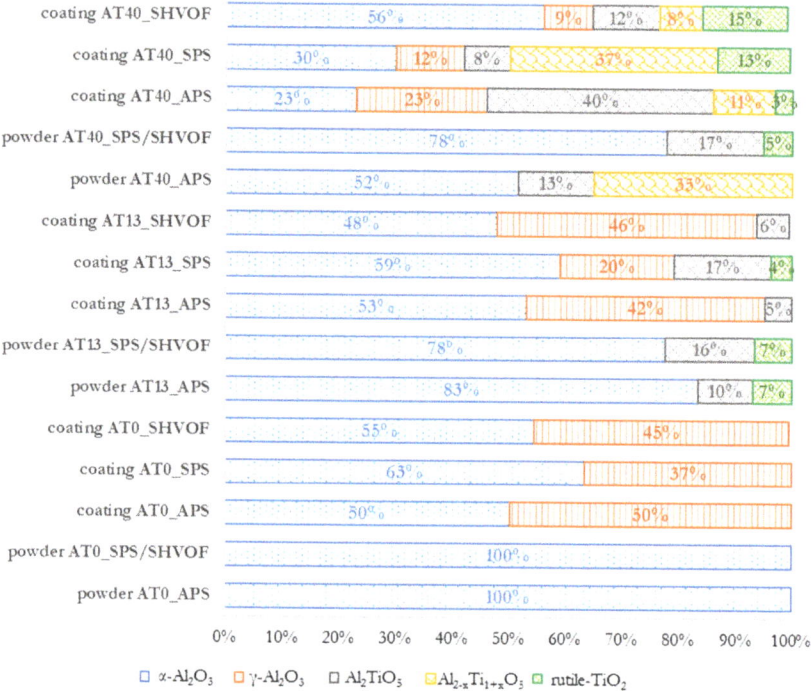

Figure 10. Quantitative estimation of the phase composition in powders and coatings.

3.3.2. APS Coatings

In AT0 APS coatings, the phase change covered the transformation of α-Al$_2$O$_3$ into γ-Al$_2$O$_3$ (Figure 11).

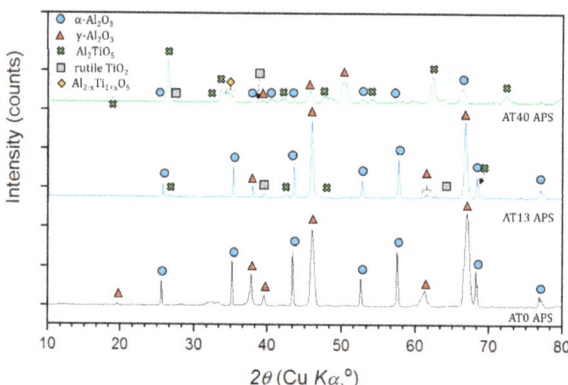

Figure 11. XRD patterns of APS-sprayed coatings.

The presence of γ-Al$_2$O$_3$ in conventional plasma sprayed coatings was the result of the rapid heating and cooling of molten powder particles. It is assumed that due to a lower activation energy of γ-Al$_2$O$_3$, its formation was favored in comparison with α-Al$_2$O$_3$ [48–52]. Moreover, the presence of α-Al$_2$O$_3$ (50 vol.%) was caused by the incomplete melting of powders in the plasma jet, as confirmed by microstructural studies. In the literature, the content of the α-Al$_2$O$_3$ phase in AT0 APS coatings is reported in a wide range, starting from 4 vol.% [53], 15 vol.% [42] up to 35 vol.% [54]. It is commonly agreed that the amount of preserved α-Al_2O_3 phase is influenced both by the characteristics of the feedstock material as well as by the spray process parameters.

In AT13 APS coatings, one of the main changes was (similarly to AT0 APS) the transformation of α-Al_2O_3 (initially 83 vol.%, after spraying 53 vol.%) into γ-Al_2O_3 (42 vol.% after spraying). This change is usually observed in the studies related to the Al_2O_3-TiO_2 system [54–57]. Compared to AT0 APS, the phase changes in AT13 APS coatings is more complex, due to the presence of tialite Al_2TiO_5 in the feedstock material. It was observed that the Al_2TiO_5 phase was reduced to half of its original content (from 10 vol.% to 5 vol.%). As explained by Vicent et al. [24], the Al_2O_3-TiO_2 feedstock in the form of micrometer-sized powder, tends to transform to Al_2TiO_5 less intensively than the suspension. Moreover, in the coating, the peaks of rutile-TiO_2 were not identified. This could result from the use of Ar/H_2 plasma-forming gases for APS spraying. It might lead to the inhomogeneous distribution of oxygen in the coating and reduction of phases derived from TiO_2 [58].

Similarly, the phase composition of the AT40 APS coating showed significant differences from the composition of the powder that was used for spraying. The major phase in this case, as expected according to the Al_2O_3-TiO_2 phase diagram, was tialite Al_2TiO_5 (40 vol.%) [59]. Under equilibrium conditions, with the chemical composition of AT40, there is only a small amount of α-Al_2O_3, together with Al_2TiO_5 [60]. Tialite phase was formed as a result of the reaction between Al_2O_3 and TiO_2 particles in the plasma jet. In AT40 APS coating the peaks of Al_2TiO_5 were obviously more intensive than in AT13 APS (Figure 11). However, this was not only due to higher TiO_2 content, but it was influenced also by the smaller size of agglomerated AT40 powder particles. It was confirmed in different studies [60–63] that the fine size of powder particles (and therefore, the larger specific surface area of particles) promotes the formation of the Al_2TiO_5 phase. Moreover, in AT40 APS coating, α-Al_2O_3 phase present in the raw material (52 vol.%) was transformed into γ-Al_2O_3 (23 vol.% of α-Al_2O_3 and 23 vol.% of γ-Al_2O_3 in the coating). This type of transformation (observed also for AT0 APS and AT13 APS coatings) is typical for conventional thermal spraying. Additionally, it was assumed that the solid solution of the $Al_{2-x}Ti_{1+x}O_5$ phase (rich in alumina) was oxidized and decomposed to Al_2TiO_5, which was also observed by Richter et al. [58].

As already described, the XRD patterns of APS-sprayed coatings showed that all powders underwent significant phase transformations during spraying. In the case of AT0 and AT13, the decrease in α-Al_2O_3, which partially transformed into γ-Al_2O_3 was the most significant change. According to the results, with the increased content of TiO_2 (AT40), the number of phases related to pure Al_2O_3 (α-Al_2O_3 and γ-Al_2O_3) was considerably reduced (Figure 12). This was accompanied by the increase in phases derived from TiO_2.

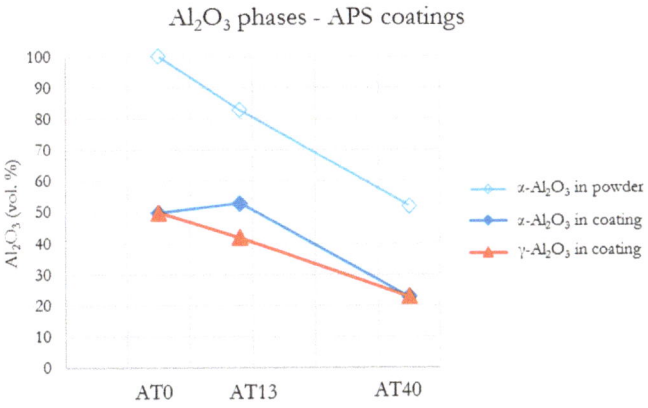

Figure 12. Al_2O_3 phases content in APS coatings.

3.3.3. SPS Coatings

According to the XRD patterns given in Figure 13, pure Al$_2$O$_3$ SPS coatings contained both a stable and metastable Al$_2$O$_3$ phases.

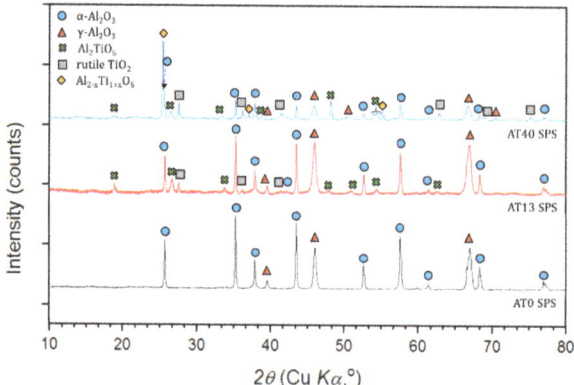

Figure 13. XRD patterns of SPS sprayed coatings.

The presence of α-Al$_2$O$_3$ (63 vol.%) and γ-Al$_2$O$_3$ (37 vol.%) phases in AT0 SPS coating was mainly caused by: (i) fraction of nonmelted particles in the final coating (Figure 4) and (ii) α→γ phase change during coating deposition, i.e., some part of fine powder particles were rapidly melted, cooled and crystallized, respectively [48]. Furthermore, the AT0 SPS sample had the highest content of α-Al$_2$O$_3$ among all AT0 coatings. This was probably influenced by the small size of powder particles and radial suspension injection. In such configurations, the finest fraction of powder tends to follow the external and relatively cold regions of plasma. This influences the heat-history of particles, causes some difficulties in particle melting and has an effect on particle impact on the substrate (this may also explain the cauliflower topography of the coating in Figure 3). The fact that the spraying was carried out with the use of a cascaded gun, with a spray distance of 80 mm could be relevant in this case as well. As for suspension-based spraying, it was rather a long spray distance (usually it is of about 40, similarly like here [64]). Thus, the velocity, trajectory and temperature distribution of the sprayed powder particles could be influenced as well. However, in general, the heat gradient and cooling rate of deposited particles should not be that significant in such cases, so this would limit the transformation into γ-Al$_2$O$_3$. Comparing the obtained results with the works of other authors, it should be pointed that the content of α-Al$_2$O$_3$ in AT0 SPS coatings is reported in a varied range, between 18 vol.% [65], 25 vol.% [66] and even up to 65-77 vol.% [32].

Considering the phase composition of AT13 SPS coating, it was observed that a suspension-based coating was of higher α-Al$_2$O$_3$ content than the APS one. SEM observations showed that the retention of α-Al$_2$O$_3$ may be the result of nonfully melted powder particles in the coating structure but also of relatively slow cooling, solidification and crystallization of splats, similarly as discussed in the case of AT0 SPS. Moreover, it is suggested that most of the initial rutile-TiO$_2$ (7 vol.%) reacted with alumina during spraying, which led to the formation of Al$_2$TiO$_5$ (17 vol.%). Such a transformation was also observed in the works of other authors investigating SPS coatings with different TiO$_2$ contents [67].

In the case of AT40 SPS coating, the intermediate phase Al$_{2-x}$Ti$_{1+x}$O$_5$ (37 vol.% in the coating) was formed as a product of Al$_2$O$_3$ (78 vol.% α-Al$_2$O$_3$) with the tialite Al$_2$TiO$_5$ reaction. It is assumed that the solid solution of this phase is formed at the intermediate stage of the Al$_2$TiO$_5$ decomposition to the form of Al$_2$O$_3$ and TiO$_2$ [60]. This is confirmed also by the presence of Al$_2$O$_3$ phases, identified in the coating (30 vol.% of α-Al$_2$O$_3$ and 12 vol.% of γ-Al$_2$O$_3$).

Similarly, as in the case of APS coatings, the results showed that with an increased TiO_2 content, SPS coatings were characterized by decreased content of α-Al_2O_3 and γ-Al_2O_3 (Figure 14). Moreover, phases in the form of: rutile-TiO_2, tialite Al_2TiO_5 and $Al_{2-x}Ti_{1+x}O_5$ were identified in SPS coatings. These findings were consistent with the observations of other authors working on Al_2O_3-TiO_2 coating produced by using submicrometer- and nanometer-sized powders [23,24,67–69].

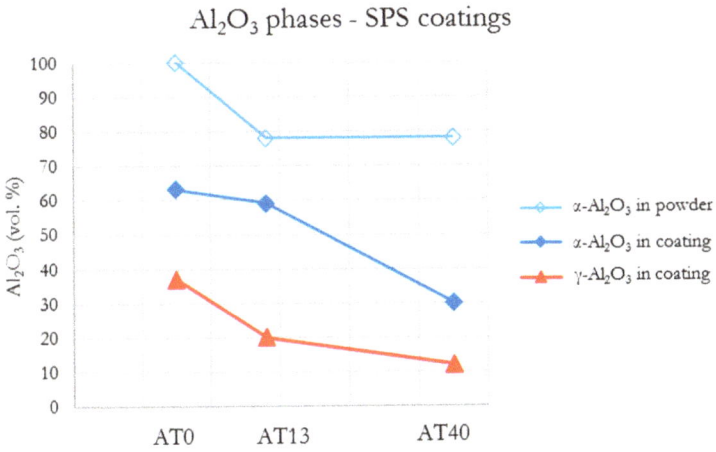

Figure 14. Al_2O_3 phases content in SPS coatings.

3.3.4. S-HVOF Coatings

In comparison with the initial AT0 powder (which consisted completely of stable α-Al_2O_3), AT0 S-HVOF coatings contained α-Al_2O_3 (55 vol.%) and γ-Al_2O_3 (45 vol.%), which indicated the transformation of α-Al_2O_3 into γ-Al_2O_3 (Figure 15).

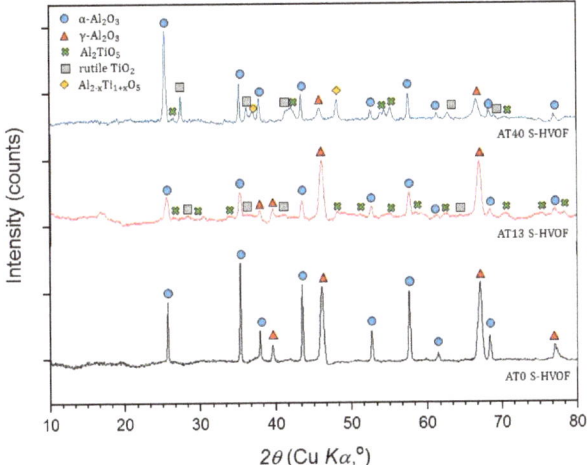

Figure 15. XRD patterns of S-HVOF sprayed coatings.

So far, in the literature, a wide range of α-Al_2O_3/γ-Al_2O_3 ratios in AT0 S-HVOF coatings was published: 2.8 vol.% [10], 5 vol.% [18], 19–73 vol.% [32]. It is considered that α-Al_2O_3 in the studied coatings did not result (at least partially) from the presence of nonmelted powders [23], as the loosely bonded powder particles were not observed in the coating. The microscopic studies revealed lamellar characteristics with well-flattened splats, as discussed in previous paragraphs. It might be possible that, as already suggested by Toma et al. [32], the substrate interpass temperature during spraying (250–350 °C) had a positive impact on the retention of α-Al_2O_3. This sufficiently limited the cooling rate of splats and preserved the α-Al_2O_3 phase in the coating.

AT13 and AT40 S-HVOF coatings showed different mechanisms of phase changes, mainly because of differences in TiO_2 content. In AT13 coatings, a higher number of phases derived from Al_2O_3 was identified (48 vol.% of α-Al_2O_3 and 46 vol.% of γ-Al_2O_3). However, the use of AT40 powder allowed the retention of more of the desired α-Al_2O_3 phase (56 vol.%) when compared with AT13. At the same time, the content of γ-Al_2O_3 in AT40 S-HVOF coatings was very low (less than 9 vol.%). According to SEM observations, the splats in AT40 S-HVOF coating were very well melted, and therefore, the identified α-Al_2O_3 was assumed to have remained as a result of relatively slow solidification, as discussed already.

Moreover, the amount of Al_2TiO_5 in AT13 S-HVOF coatings was reduced when compared to the composition of starting powder. It was twice lower (6 vol.%) than in the case of AT40 S-HVOF coating (12 vol.%). Additionally, in AT40 S-HVOF coatings (similarly as in the case of SPS), the presence of intermediate phase $Al_{2-x}Ti_{1+x}O_5$ was observed (8 vol.%).

Quantitative analysis of S-HVOF coatings showed that with higher TiO_2 content, the presence of rutile-TiO_2 (AT13, AT40), Al_2TiO_5 (AT13 and AT40), $Al_{2-x}Ti_{1+x}O_5$ (AT40) was identified (Figure 10). Moreover, based on XRD diffractograms, it is expected that in all S-HVOF coatings the amorphous phases existed, probably due to still relatively intensive heating/cooling conditions [54]. A trend in decreasing the pure Al_2O_3 phases content along with an increase in TiO_2 was still observed but it was not that obvious as in the case of APS and SPS coatings, especially for AT40 (Figure 16). Unfortunately, it was not possible to truly compare the presented results to the literature, as there are no papers concerning such types of coatings yet.

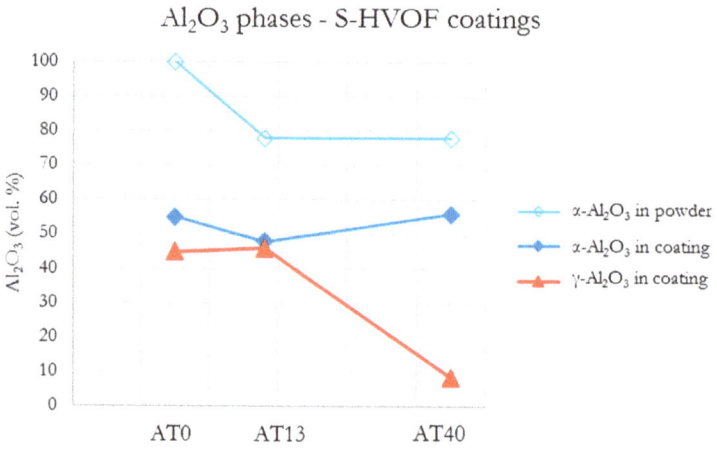

Figure 16. Al_2O_3 phases content in S-HVOF coatings.

The results showed that not only the chemical composition but also the spraying method had an influence on the phase composition of the obtained coatings. Significant differences were observed for α-Al_2O_3 and γ-Al_2O_3. Among all spraying techniques, the highest content of α-Al_2O_3 was obtained for the following coatings: (i) for pure Al_2O_3 in the case of SPS coatings (63 vol.%), (ii) for Al_2O_3 + 13 wt.% TiO_2 in the case of SPS coatings (59 vol.%), (iii) for Al_2O_3 + 40 wt.% TiO_2 in the case of S-HVOF coating (56 vol.%). With regard to intermediary and rutile-TiO_2 phases, the differences were observed especially between coatings obtained from powders and liquid feedstocks (Figure 17), i.e., in the AT40 APS coating, the content of tialite Al_2TiO_5 was several times higher than in AT40 SPS and AT40 S-HVOF coatings.

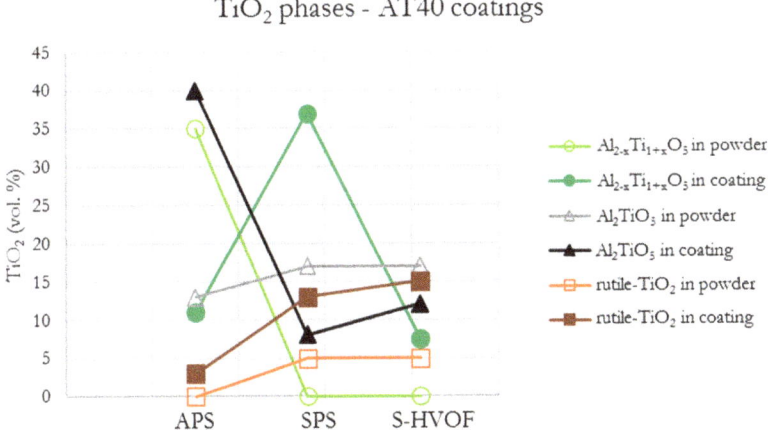

Figure 17. Content of TiO_2 and intermediary phases in AT40 coatings.

4. Conclusions

In the presented study, Al_2O_3, Al_2O_3 + 13 wt.% TiO_2 and Al_2O_3 + 40 wt.% TiO_2 coatings were successfully deposited by using APS, SPS and S-HVOF thermal spray methods. Water-based suspensions sprayed by SPS and S-HVOF allowed producing of dense and more homogeneous coatings than those obtained by conventional APS. S-HVOF coatings were characterized by fine porosity and smooth coating surface, while SPS coatings exhibited more porous microstructure, but with still evenly distributed pores. It was also found that cascade the KK plasma gun and S-HVOF Top Gun enabled the deposition of suspension coatings at comparable stand-off distances as in conventional APS and HVOF processes, which is important when, for example, coating parts with complex geometry.

The alumina-titania suspensions were formulated here by using preprocessed commercially available micrometer-sized powders. High-energy ball milling may be easily used to obtain such submicrometer-sized Al_2O_3–TiO_2 powders but the mechanical treatment introduced initial phase changes to the feedstock material, even if it was in agglomerated form. Then, the serious phase transformations occurred during spraying, depending on, e.g., the Al_2O_3/TiO_2 ratio, deposition method, spray parameters, etc. In general, in the case of AT0 and AT13, the most intensively observed was the decrease in α-Al_2O_3, which partially transformed into γ-Al_2O_3. The increased content of TiO_2 (AT40), caused the decrease in vol.% of Al_2O_3 (both α-Al_2O_3 and γ-Al_2O_3) and was accompanied by the increase in phases derived from TiO_2 (Al_2TiO_5 and $Al_{2-x}Ti_{1+x}O_5$). The results confirmed also that (i) α-Al_2O_3 or (ii) α-Al_2O_3 with rutile-TiO_2 may be preserved more easily in AT0, AT13 and AT40 coatings made by SPS and S-HVOF. However, the phase analysis is a complex task and a more detailed analysis will be carried out and presented in future works.

Author Contributions: Conceptualization, M.M., F.-L.T., L.L. and P.S.; methodology, F.-L.T. and L.L.; validation, F.-L.T., L.L. and P.S; investigation, M.M., L.L.; writing—original draft preparation, M.M.; writing—L.L., P.S. and F.-L.T.; supervision, M.B. and A.A.; funding acquisition, M.M. All authors have read and agreed to the published version of the manuscript.

Funding: This research was funded by scholarship of DAAD Foundation (Deutscher Akademischer Austauschdienst), Project No. 57378443.

Acknowledgments: The authors acknowledge the colleagues from Fraunhofer IWS: Oliver Kunze, Martin Köhler and Stefan Scheitz for support during spraying of the coatings, Irina Shakhverdova and Beate Wolf for metallographic preparation of suspension sprayed samples.

Conflicts of Interest: The authors declare no conflict of interest. The funders had no role in the design of the study; in the collection, analyses, or interpretation of data; in the writing of the manuscript, or in the decision to publish the results.

References

1. Fauchais, P.; Montavon, G. Latest Developments in Suspension and Liquid Precursor Thermal Spraying. *J. Therm. Spray Technol.* **2010**, *19*, 226–239. [CrossRef]
2. Killinger, A.; Gadow, R.; Mauer, G.; Guignard, A.; Vaßen, R.; Stöver, D. Review of New Developments in Suspension and Solution Precursor Thermal Spray Processes. *J. Therm. Spray Technol.* **2011**, *20*, 677. [CrossRef]
3. Gitzhofer, F.; Bouyer, E.; Boulos, M.I. Suspension Plasma Spray. Patent no. US5609921 A, 11 March 1997.
4. Gadow, R.P.D.; Killinger, A.D.; Kuhn, M.; Martinez, D.L. Verfahren und Vorrichtung zum thermischen Spritzen von Suspensionen. Patent no. DE102005038453A1, 9 June 2011.
5. Toma, F.-L.; Berger, L.-M.; Stahr, C.C.; Naumann, T.; Langner, S. Thermally Sprayed Al2O3 Coatings Having a High Content of Corundum without any Property-Reducing Additives, and Method for the Production Thereof. Patent no. US8318261 B2, 27 November 2012.
6. Potthoff, A.; Kratzsch, R.; Barbosa, M.; Kulissa, N.; Kunze, O.; Toma, F.-L. Development and Application of Binary Suspensions in the Ternary System $Cr_2O_3/TiO_2/Al_2O_3$ for S-HVOF Spraying. *J. Therm. Spray Technol.* **2018**, *27*, 710–717. [CrossRef]
7. Toma, F.-L.; Berger, L.-M.; Naumann, T.; Langner, S. Microstructures of nanostructured ceramic coatings obtained by suspension thermal spraying. *Surf. Coat. Technol.* **2008**, *202*, 4343–4348. [CrossRef]
8. Gadow, R.; Killinger, A.; Rauch, J. New results in High Velocity Suspension Flame Spraying (HVSFS). *Surf. Coat. Technol.* **2008**, *202*, 4329–4336. [CrossRef]
9. Toma, F.-L.; Stahr, C.C.; Berger, L.-M.; Saaro, S.; Herrmann, M.; Deska, D.; Michael, G. Corrosion Resistance of APS- and HVOF-Sprayed Coatings in the Al_2O_3-TiO_2 System. *J. Therm. Spray Technol.* **2010**, *19*, 137–147. [CrossRef]
10. Bolelli, G.; Rauch, J.; Cannillo, V.; Killinger, A.; Lusvarghi, L.; Gadow, R. Microstructural and Tribological Investigation of High-Velocity Suspension Flame Sprayed (HVSFS) Al_2O_3 Coatings. *J. Therm. Spray Technol.* **2008**, *18*, 35. [CrossRef]
11. Fauchais, P.; Etchart-Salas, R.; Rat, V.; Coudert, J.F.; Caron, N.; Wittmann-Ténèze, K. Parameters Controlling Liquid Plasma Spraying: Solutions, Sols, or Suspensions. *J. Therm. Spray Technol.* **2008**, *17*, 31–59. [CrossRef]
12. Toma, F.-L.; Potthoff, A.; Barbosa, M. Microstructural Characteristics and Performances of Cr_2O_3 and Cr_2O_3-15%TiO_2 S-HVOF Coatings Obtained from Water-Based Suspensions. *J. Therm. Spray Technol.* **2018**, *27*, 344–357. [CrossRef]
13. Tingaud, O.; Bertrand, P.; Bertrand, G. Microstructure and tribological behavior of suspension plasma sprayed Al_2O_3 and Al_2O_3–YSZ composite coatings. *Surf. Coat. Technol.* **2010**, *205*, 1004–1008. [CrossRef]
14. Kozerski, S.; Toma, F.-L.; Pawlowski, L.; Leupolt, B.; Latka, L.; Berger, L.-M. Suspension plasma sprayed TiO_2 coatings using different injectors and their photocatalytic properties. *Surf. Coat. Technol.* **2010**, *205*, 980–986. [CrossRef]
15. Bolelli, G.; Bonferroni, B.; Cannillo, V.; Gadow, R.; Killinger, A.; Lusvarghi, L.; Rauch, J.; Stiegler, N. Wear behaviour of high velocity suspension flame sprayed (HVSFS) Al_2O_3 coatings produced using micron- and nano-sized powder suspensions. *Surf. Coat. Technol.* **2010**, *204*, 2657–2668. [CrossRef]
16. Bolelli, G.; Cannillo, V.; Gadow, R.; Killinger, A.; Lusvarghi, L.; Manfredini, T.; Müller, P. Properties of Al_2O_3 coatings by High Velocity Suspension Flame Spraying (HVSFS): Effects of injection systems and torch design. *Surf. Coat. Technol.* **2015**, *270*, 175–189. [CrossRef]
17. Rauch, J.; Bolelli, G.; Killinger, A.; Gadow, R.; Cannillo, V.; Lusvarghi, L. Advances in High Velocity Suspension Flame Spraying (HVSFS). *Surf. Coat. Technol.* **2009**, *203*, 2131–2138. [CrossRef]
18. Murray, J.W.; Ang, A.S.M.; Pala, Z.; Shaw, E.C.; Hussain, T. Suspension High Velocity Oxy-Fuel (SHVOF)-Sprayed Alumina Coatings: Microstructure, Nanoindentation and Wear. *J. Therm. Spray Technol.* **2016**, *25*, 1700–1710. [CrossRef]

19. Darut, G.; Ben-Ettouil, F.; Denoirjean, A.; Montavon, G.; Ageorges, H.; Fauchais, P. Dry Sliding Behavior of Sub-Micrometer-Sized Suspension Plasma Sprayed Ceramic Oxide Coatings. *J. Therm. Spray Technol.* **2010**, *19*, 275–285. [CrossRef]
20. Goel, S.; Björklund, S.; Curry, N.; Wiklund, U.; Joshi, S. Axial suspension plasma spraying of Al_2O_3 coatings for superior tribological properties. *Surf. Coat. Technol.* **2017**, *315*, 80–87. [CrossRef]
21. Owoseni, T.A.; Murray, J.W.; Pala, Z.; Lester, E.H.; Grant, D.M.; Hussain, T. Suspension high velocity oxy-fuel (SHVOF) spray of delta-theta alumina suspension: Phase transformation and tribology. *Surf. Coat. Technol.* **2019**, *371*, 97–106. [CrossRef]
22. Bolelli, G.; Cannillo, V.; Gadow, R.; Killinger, A.; Lusvarghi, L.; Rauch, J.; Romagnoli, M. Effect of the suspension composition on the microstructural properties of high velocity suspension flame sprayed (HVSFS) Al_2O_3 coatings. *Surf. Coat. Technol.* **2010**, *204*, 1163–1179. [CrossRef]
23. Darut, G.; Klyatskina, E.; Valette, S.; Carles, P.; Denoirjean, A.; Montavon, G.; Ageorges, H.; Segovia, F.; Salvador, M. Architecture and phases composition of suspension plasma sprayed alumina-titania sub-micrometer-sized coatings. *Mater. Lett.* **2012**, *67*, 241–244. [CrossRef]
24. Vicent, M.; Bannier, E.; Carpio, P.; Rayón, E.; Benavente, R.; Salvador, M.D.; Sánchez, E. Effect of the initial particle size distribution on the properties of suspension plasma sprayed Al_2O_3–TiO_2 coatings. *Surf. Coat. Technol.* **2015**, *268*, 209–215. [CrossRef]
25. ASTM B822—17. *Standard Test Method for Particle Size Distribution of Metal Powders and Related Compounds by Light Scattering*; ASTM International: West Conshohocken, PA, USA, 2017.
26. Toma, F.-L.; Berger, L.-M.; Scheitz, S.; Langner, S.; Rödel, C.; Potthoff, A.; Sauchuk, V.; Kusnezoff, M. Comparison of the Microstructural Characteristics and Electrical Properties of Thermally Sprayed Al_2O_3 Coatings from Aqueous Suspensions and Feedstock Powders. *J. Therm. Spray Technol.* **2012**, *21*, 480–488. [CrossRef]
27. Łatka, L.; Niemiec, A.; Michalak, M.; Sokołowski, P. Tribological properties of Al_2O_3 + TiO_2 coatings manufactured by plasma spraying. *Tribology* **2019**, *1*, 19–24. [CrossRef]
28. Łatka, L.; Szala, M.; Michalak, M.; Pałka, T. Impact of Atmospheric Plasma Spray Parameters on Cavitation Erosion Resistance of Al_2O_3 –13% TiO_2 Coatings. *Acta Phys. Pol. A* **2019**, *136*, 342–347. [CrossRef]
29. ASTM E2109-01(2014). *Standard Test Methods for Determining Area Percentage Porosity in Thermal Sprayed Coatings*; ASTM International: West Conshohocken, PA, USA, 2014.
30. Prevéy, P.S. X-ray diffraction characterization of crystallinity and phase composition in plasma-sprayed hydroxyapatite coatings. *J. Therm. Spray Technol.* **2000**, *9*, 369–376. [CrossRef]
31. Marcinauskas, L.; Valatkevičius, P. The effect of plasma torch power on the microstructure and phase composition of alumina coatings. *Mater. Sci.* **2010**, *28*, 451–458.
32. Toma, F.-L.; Berger, L.-M.; Stahr, C.C.; Naumann, T.; Langner, S. Microstructures and Functional Properties of Suspension-Sprayed Al_2O_3 and TiO_2 Coatings: An Overview. *J. Therm. Spray Technol.* **2010**, *19*, 262–274. [CrossRef]
33. Carpio, P.; Salvador, M.D.; Borrell, A.; Sánchez, E.; Moreno, R. Alumina-zirconia coatings obtained by suspension plasma spraying from highly concentrated aqueous suspensions. *Surf. Coat. Technol.* **2016**, *307*, 713–719. [CrossRef]
34. Chidambaram Seshadri, R.; Sampath, S. Characteristics of Conventional and Cascaded Arc Plasma Spray-Deposited Ceramic Under Standard and High-Throughput Conditions. *J. Therm. Spray Technol.* **2019**, *28*, 690–705. [CrossRef]
35. Fazilleau, J.; Delbos, C.; Rat, V.; Coudert, J.F.; Fauchais, P.; Pateyron, B. Phenomena Involved in Suspension Plasma Spraying Part 1: Suspension Injection and Behavior. *Plasma Chem. Plasma Process.* **2006**, *26*, 371–391. [CrossRef]
36. Killinger, A. 4—Status and future trends in suspension spray techniques. In *Future Development of Thermal Spray Coatings*; Espallargas, N., Ed.; Woodhead Publishing: Sawston, UK, 2015; pp. 81–122. ISBN 978-0-85709-769-9.
37. Steeper, T.J.; Varacalle, D.J.; Wilson, G.C.; Riggs, W.L.; Rotolico, A.J.; Nerz, J. A design of experiment study of plasma-sprayed alumina-titania coatings. *JTST* **1993**, *2*, 251–256. [CrossRef]
38. Habib, K.A.; Saura, J.J.; Ferrer, C.; Damra, M.S.; Giménez, E.; Cabedo, L. Comparison of flame sprayed Al_2O_3/TiO_2 coatings: Their microstructure, mechanical properties and tribology behavior. *Surf. Coat. Technol.* **2006**, *201*, 1436–1443. [CrossRef]

39. Łatka, L. Thermal Barrier Coatings Manufactured by Suspension Plasma Spraying—A Review. *Adv. Mater. Sci.* **2018**, *18*, 95–117. [CrossRef]
40. Michalak, M.; Łatka, L.; Szymczyk, P.; Sokołowski, P. Computational image analysis of Suspension Plasma Sprayed YSZ coatings. In *ITM Web of Conferences*; EDP Sciences: Les Ulis, France, 2017; Volume 15, p. 06004. [CrossRef]
41. Müller, P.; Killinger, A.; Gadow, R. Comparison Between High-Velocity Suspension Flame Spraying and Suspension Plasma Spraying of Alumina. *J. Therm. Spray Technol.* **2012**, *21*, 1120–1127. [CrossRef]
42. Franco, D.; Ageorges, H.; Lopez, E.; Vargas, F. Tribological performance at high temperatures of alumina coatings applied by plasma spraying process onto a refractory material. *Surf. Coat. Technol.* **2019**, *371*, 276–286. [CrossRef]
43. Zuluaga, C.M.S. Mechanical Behavior of Al_2O_3-13%TiO_2 Ceramic Coating at Elevated Temperature. Master's Thesis, Universidad Nacional de Colombia—Sede Medellín, Antioquia, Colombia, 2016.
44. Barth, N.; Schilde, C.; Kwade, A. Influence of Particle Size Distribution on Micromechanical Properties of thin Nanoparticulate Coatings—ScienceDirect. *Phys. Procedia* **2013**, *40*, 9–18. [CrossRef]
45. Yang, G.-J.; Suo, X. (Eds.) *Advanced Nanomaterials and Coatings by Thermal Spray: Multi-Dimensional Design of Micro-Nano Thermal Spray Coatings*, 1st ed.; Elsevier: Amsterdam, The Netherlands, 2019; ISBN 978-0-12-813870-0.
46. Bégin-Colin, S.; Gadalla, A.; Caër, G.; Humbert, O.; Thomas, F.; Barres, O.; Villiéras, F.; Toma, F.L.; Bertrand, G.; Zahraa, O.; et al. On the Origin of the Decay of the Photocatalytic Activity of TiO_2 Powders Ground at High Energy. *J. Phys. Chem. C* **2009**, *113*, 16589–16602. [CrossRef]
47. Bégin-Colin, S.; Girot, T.; Le Caër, G.; Mocellin, A. Kinetics and Mechanisms of Phase Transformations Induced by Ball-Milling in Anatase TiO_2. *J. Solid State Chem.* **2000**, *149*, 41–48. [CrossRef]
48. Goberman, D.; Sohn, Y.H.; Shaw, L.; Jordan, E.; Gell, M. Microstructure development of Al_2O_3–13wt.%TiO_2 plasma sprayed coatings derived from nanocrystalline powders. *Acta Mater.* **2002**, *50*, 1141–1152. [CrossRef]
49. McPherson, R. Formation of metastable phases in flame- and plasma-prepared alumina. *J. Mater. Sci.* **1973**, *8*, 851–858. [CrossRef]
50. Rico, A.; Rodriguez, J.; Otero, E.; Zeng, P.; Rainforth, W.M. Wear behaviour of nanostructured alumina-titania coatings deposited by atmospheric plasma spray. *Wear* **2009**, *267*, 1191–1197. [CrossRef]
51. Dachille, F.; Simons, P.Y.; Roy, R. Pressure-temperature studies of anatase, brookite, rutile and TiO_2-II. *Am. Mineral.* **1968**, *53*, 1929–1939.
52. Shaw, L.L.; Goberman, D.; Ren, R.; Gell, M.; Jiang, S.; Wang, Y.; Xiao, T.D.; Strutt, P.R. The dependency of microstructure and properties of nanostructured coatings on plasma spray conditions. *Surf. Coat. Technol.* **2000**, *130*, 1–8. [CrossRef]
53. Jia, S.; Zou, Y.; Xu, J.; Wang, J.; Yu, L. Effect of TiO_2 content on properties of Al_2O_3 thermal barrier coatings by plasma spraying. *Trans. Nonferr. Met. Soc. China* **2015**, *25*, 175–183. [CrossRef]
54. Dejang, N.; Watcharapasorn, A.; Wirojupatump, S.; Niranatlumpong, P.; Jiansirisomboon, S. Fabrication and properties of plasma-sprayed Al_2O_3/TiO_2 composite coatings: A role of nano-sized TiO_2 addition. *Surf. Coat. Technol.* **2010**, *204*, 1651–1657. [CrossRef]
55. Stengl, V.; Ageorges, H.; Ctibor, P.; Murafa, N. Atmospheric plasma sprayed (APS) coatings of Al_2O_3-TiO_2 system for photocatalytic application. *Photochem. Photobiol. Sci.* **2009**, *8*, 733–738. [CrossRef] [PubMed]
56. Vijay, M.; Selvarajan, V.; Yugeswaran, S.; Ananthapadmanabhan, P.V.; Sreekumar, K.P. Effect of Spraying Parameters on Deposition Efficiency and Wear Behavior of Plasma Sprayed Alumina-Titania Composite Coatings. *Plasma Sci. Technol.* **2009**, *11*, 666–673. [CrossRef]
57. Vicent, M.; Bannier, E.; Moreno, R.; Salvador, M.D.; Sánchez, E. Atmospheric plasma spraying coatings from alumina-titania feedstock comprising bimodal particle size distributions. *J. Eur. Ceram. Soc.* **2013**, *33*, 3313–3324. [CrossRef]
58. Richter, A.; Berger, L.-M.; Sohn, Y.J.; Conze, S.; Sempf, K.; Vaßen, R. Impact of Al_2O_3-40 wt.% TiO_2 feedstock powder characteristics on the sprayability, microstructure and mechanical properties of plasma sprayed coatings. *J. Eur. Ceram. Soc.* **2019**, *39*, 5391–5402. [CrossRef]
59. Goldberg, D. Contribution to study of systems formed by alumina and some oxides of trivalent and tetravalent metals especially titanium oxide. *Rev. Intern. Hautes Temp. Refractaires* **1968**, *5*, 181–182.

60. Berger, L.-M.; Sempf, K.; Sohn, Y.J.; Vaßen, R. Influence of Feedstock Powder Modification by Heat Treatments on the Properties of APS-Sprayed Al_2O_3-40% TiO_2 Coatings. *J. Therm. Spray Technol.* **2018**, *27*, 654–666. [CrossRef]
61. Islak, S.; Buytoz, S.; Ersöz, E.; Orhan, N.; Stokes, J.; Saleem Hashmı, M.; Somunkıran, I.; Tosun, N. Effect on microstructure of TiO_2 rate in Al_2O_3-TiO_2 composite coating produced using plasma spray method. *Optoelectron. Adv. Mater. Rapid Commun.* **2012**, *6*, 844–849.
62. Freudenberg, F. Etude de la réaction à l'état solide $Al_2O_3 + TiO_2 \rightarrow Al_2TiO_5$: Observation des structures. Ph.D. Thesis, The École Polytechnique Fédérale de Lausanne EPFL, Lausanne, Switzerland, 1988; p. 709.
63. Hoffmann, S.; Norberg, S.T.; Yoshimura, M. Melt synthesis of Al_2TiO_5 containing composites and reinvestigation of the phase diagram Al_2O_3–TiO_2 by powder X-ray diffraction. *J. Electroceram.* **2006**, *16*, 327–330. [CrossRef]
64. Sokołowski, P.; Kozerski, S.; Pawłowski, L.; Ambroziak, A. The key process parameters influencing formation of columnar microstructure in suspension plasma sprayed zirconia coatings. *Surf. Coat. Technol.* **2014**, *260*, 97–106. [CrossRef]
65. Stahr, C.C.; Saaro, S.; Berger, L.-M.; Dubsky, J.; Neufuss, K.; Herrmann, M. Dependence of the Stabilization of α-Alumina on the Spray Process. *J. Therm. Spray Technol.* **2007**, *16*, 822–830. [CrossRef]
66. Tesar, T.; Musalek, R.; Medricky, J.; Kotlan, J.; Lukac, F.; Pala, Z.; Ctibor, P.; Chraska, T.; Houdkova, S.; Rimal, V.; et al. Development of suspension plasma sprayed alumina coatings with high enthalpy plasma torch. *Surf. Coat. Technol.* **2017**, *325*, 277–288. [CrossRef]
67. Bannier, E.; Vicent, M.; Rayón, E.; Benavente, R.; Salvador, M.D.; Sánchez, E. Effect of TiO_2 addition on the microstructure and nanomechanical properties of Al_2O_3 Suspension Plasma Sprayed coatings. *Appl. Surf. Sci.* **2014**, *316*, 141–146. [CrossRef]
68. Klyatskina, E.; Rayón, E.; Darut, G.; Salvador, M.D.; Sánchez, E.; Montavon, G. A study of the influence of TiO_2 addition in Al_2O_3 coatings sprayed by suspension plasma spray. *Surf. Coat. Technol.* **2015**, *278*, 25–29. [CrossRef]
69. Klyatskina, E.; Espinosa-Fernández, L.; Darut, G.; Segovia, F.; Salvador, M.D.; Montavon, G.; Agorges, H. Sliding Wear Behavior of Al_2O_3–TiO_2 Coatings Fabricated by the Suspension Plasma Spraying Technique. *Tribol. Lett.* **2015**, *59*, 8. [CrossRef]

© 2020 by the authors. Licensee MDPI, Basel, Switzerland. This article is an open access article distributed under the terms and conditions of the Creative Commons Attribution (CC BY) license (http://creativecommons.org/licenses/by/4.0/).

Letter

Exploiting Suspension Plasma Spraying to Deposit Wear-Resistant Carbide Coatings

Satyapal Mahade [1,*], Karthik Narayan [1], Sivakumar Govindarajan [2], Stefan Björklund [1], Nicholas Curry [3] and Shrikant Joshi [1]

1. Department of Engineering Science, University West, 46132 Trollhättan, Sweden
2. International Advanced Research Center for Powder Metallurgy and New Materials, Hyderabad 500069, India
3. Treibacher Industrie AG, 9330 Althofen, Austria
* Correspondence: satyapal.mahade@hv.se

Received: 10 June 2019; Accepted: 18 July 2019; Published: 24 July 2019

Abstract: Titanium- and chromium-based carbides are attractive coating materials to impart wear resistance. Suspension plasma spraying (SPS) is a relatively new thermal spray process which has shown a facile ability to use sub-micron and nano-sized feedstock to deposit high-performance coatings. The specific novelty of this work lies in the processing of fine-sized titanium and chromium carbides (TiC and Cr_3C_2) in the form of aqueous suspensions to fabricate wear-resistant coatings by SPS. The resulting coatings were characterized by surface morphology, microstructure, phase constitution, and micro-hardness. The abrasive, erosive, and sliding wear performance of the SPS-processed TiC and Cr_3C_2 coatings was also evaluated. The results amply demonstrate that SPS is a promising route to manufacture superior wear-resistant carbide-based coatings with minimal in situ oxidation during their processing.

Keywords: titanium carbide; chromium carbide; suspension plasma spray; wear

1. Introduction:

Wear is a severe problem in a vast majority of industrial applications, leading to reduced durability of engineering components and increased frequency of replacement shutdowns. Protective coatings of carbide-based compositions have been extensively employed to enhance wear resistance, and the popular processing routes to produce them have included physical vapor deposition (PVD), chemical vapor deposition (CVD), thermal spray, etc. [1]. Among these, thermal spraying offers unique advantages such as the ability to spray thicker protective layers, faster deposition rates, and cost efficiency. Historically, atmospheric plasma spraying (APS), high velocity oxy-fuel (HVOF) spraying, and more recently high-velocity air-fuel (HVAF) spraying techniques have progressively become the thermal spray methods of choice for depositing carbide-based coatings [2]. A significant limitation of these techniques is their inability to process fine powders due to considerable feeding-related challenges [3], although employing such feedstock can potentially result in improved tribological performance of the coatings [4].

Suspension plasma spraying (SPS) is an advancement in APS, enabling spraying of fine feedstock (100 nm–5 µm diameter) [5], and has already shown promise for yielding thermal barrier coatings (TBCs) with improved performance compared to APS [6]. SPS coatings can similarly enhance wear performance due to unique microstructural features inherently associated with the process, such as the smaller splat size, fine scale porosity, low surface roughness, etc. An illustrative description of distinct microstructures resulting from the thermal spraying of different feedstock sizes, as well as their likely influence on wear performance, is shown in Figure 1. Yet, the predominant focus of SPS research so far has been on TBCs. In particular, there have been very few prior efforts to deposit pure carbide coatings by SPS, which are ideal candidates for mitigating wear. Recently, Mubarok et al. reported SPS-processed SiC coatings, which exhibited excellent carbide phase retention post spraying [7]. Berghaus et al. reported similar findings (minimal oxidation) for SPS-processed WC-Co coatings, demonstrating the capability of SPS to minimize the in-flight oxidation of carbide feedstock [8]. The reason for the minimal oxidation of carbides in SPS-processed coatings compared to APS can be attributed to the presence of a solvent (ethanol or water) which consumes a considerable part of the plasma energy during its evaporation, thus minimizing feedstock decomposition.

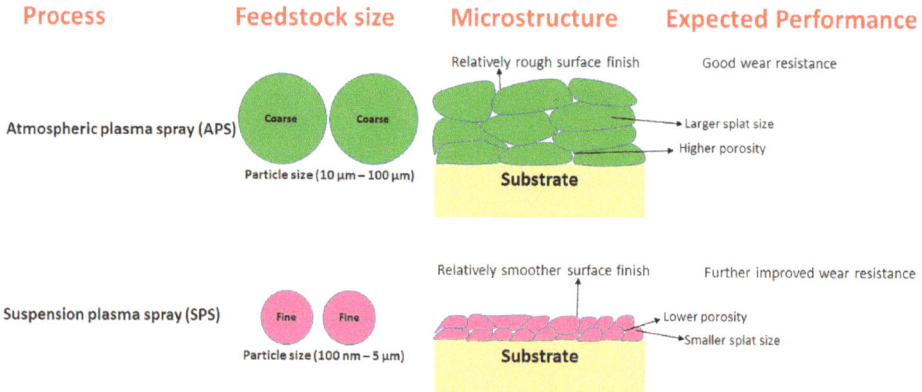

Figure 1. Schematic of comparison of feedstock–microstructure–property relationships in APS and SPS coatings.

Recognizing the above, the present study specifically deals with titanium carbide (TiC) and chromium carbide (Cr_3C_2) coatings (without any metal or alloy binder) produced by SPS. To the best of our knowledge, no prior attempts have been made to deposit TiC- and Cr_3C_2-based coatings via the SPS route. The SPS carbide coatings were comprehensively characterized using SEM, XRD, porosity, hardness, etc. The tribological behavior of the coatings under erosive, sliding, and abrasive wear modes was also evaluated.

2. Experimental Work

SSAB Domex®350LA (low-alloyed steel, SSAB AB, Stockholm, Sweden) substrates were grit blasted to provide a surface roughness of approximately 3 μm Ra prior to spraying. Two experimental water-based suspensions, comprising 40 wt.% TiC and Cr_3C_2, respectively, were produced by the Treibacher Industrie AG (Althofen, Austria) for this work. Water-based suspensions were chosen due to their higher surface tension versus ethanol, which promotes the formation of relatively denser coatings [9]. Particle size analysis of the liquid feedstock was performed using CILAS 1064 equipment (CILAS, Orleans, France). An Axial III plasma torch (Mettech Corp., Vancouver, Canada) was employed to deposit the coatings. The spray parameters used for SPS deposition of both the coatings are given in Table 1. A gas mixture comprising argon, nitrogen, and hydrogen was used as the primary gas.

Table 1. Spray parameters used for depositing TiC and Cr_3C_2 coatings.

Suspension Feed Rate (mL/min)	Nozzle Diameter (inches)	Spray Distance (mm)	Surface Speed (mm/s)	Atomizing Gas Flow Rate (L/min)	Power (kW)	Enthalpy (kJ/I)	Current (A)
40	3/8	100	100	15	111	9.7	200

The surface morphology and cross-sectional microstructure of the coatings were analyzed using a scanning electron microscope (HITACHI TM-3000, Tokyo, Japan) in back-scattered electron (BSE) mode.

Surface roughness was examined using a stylus-based profilometer (MITUTOYO SURFTEST-301, Kawasaki, Tokyo, Japan). Micro-hardness measurements were performed on the cross section of coatings (HMV-2 series, SHIMADZU Corp., Tokyo, Japan). Hardness testing was performed using a load of 980.7 mN ($HV_{0.1}$) and the dwell time was kept at 15 s. Ten independent measurements were made, and their mean and standard deviation values are reported herein. Porosity measurements were made using ImageJ software (version 1.52p, University of Wisconsin, Wisconsin, US)) [10] by considering fifteen different cross-sectional SEM micrographs at 5000× magnification. XRD analysis of the top surface of as-sprayed coatings was performed using a high-energy intensity micro X-ray diffractometer (RAPID-II-D/MAX, Rigaku Corp., Tokyo, Japan). A slow scan rate comprising a step size of 0.01° and time of 10 s per step was utilized.

The as-sprayed specimens were polished to a surface roughness <1 µm on the Ra scale prior to wear testing. Three different wear tests (erosion, sliding, and abrasion) were performed in this work. The erosion test (TR-470, DUCOM, Bengaluru, India) was performed at room temperature according to the ASTM G76-13 standard [11] at an impingement angle of 90°. Alumina of approximately 50 ± 10 µm mean particle size was used as the erodent media. During the test, the air pressure was maintained at 0.5 bar and the specimens (1-inch diameter coupons) were exposed to the erosion test for 10 min. The abrasive-wear test was performed as per ASTM G65 standard [12] using a dry abrasion test rig from DUCOM (DUCOM instruments, Bangalore, India, 2019) with 80 mesh silica sand particles at a load of 5 kg. The speed of the wheel was kept at 245 ± 5 rpm and the abrasive flow rate was kept at 350 g/min. For the sliding-wear tests, specimens of size 10 mm ×10 mm × 4 mm were cut from the coated plates. Sliding wear tests were performed as per ASTM G99 standard [13] by employing a coated pin (sintered WC-6Co) of diameter 6 mm, at a load of 5 kgf sliding against a disc rotated at 5 m/s velocity. The duration of this test was three hours and twenty minutes. In each of the above wear tests, the weight loss was measured using a high-accuracy weighing balance (Sartorius, Cubis®II, Sartorius Gmbh, Göttingen, Germany), accuracy: 0.01 mg). Three samples were used for erosion tests, while only one specimen was used for sliding- and abrasive-wear tests.

3. Results and Discussion

The particle size distribution of titanium carbide and chromium carbide feedstock is shown in Figure 2a,b respectively. The median (D_{50}) particle size of titanium carbide was approximately 2.21 µm, and the D_{90} was approximately 4.30 µm. The median (D_{50}) particle size of chromium carbide was 3.84 µm and D_{90} was 6.25 µm. The particle size of chromium carbide was only slightly larger than that of titanium carbide, which is not expected to significantly influence splat size in the deposited coatings.

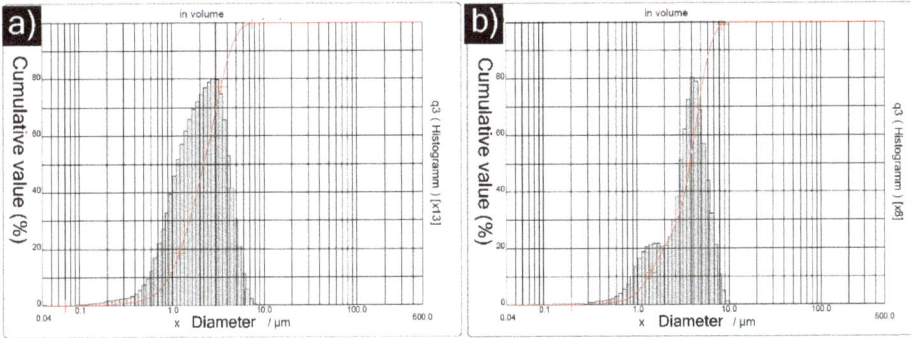

Figure 2. Particle size distribution of (**a**) TiC powder feedstock (**b**) Cr_3C_2 powder feedstock.

The SEM micrographs of TiC and Cr_3C_2 powders revealed that they were both comprised of irregularly shaped particles (see Figure 3a,b). The Cr_3C_2 particles were clearly seen to be relatively coarser than the TiC particles, confirming the accompanying particle size distribution results, according to Figure 3b.

Figure 3. SEM micrographs showing§; (**a**) feedstock TiC powder; (**b**) feedstock Cr_3C_2 powder.

The top surface views of SPS-processed TiC and Cr_3C_2 coatings in Figure 4 show fine-structured splats in the size range 3–4 μm, as compared to conventional powder-derived splats, which are at least an order of magnitude larger. The reason for the finer splat size in SPS can be directly attributed mainly to the considerably smaller particles in the feedstock. Moreover, the relatively lower droplet momentum in the case of SPS ensures that the droplets flatten to a lesser extent on impact with the substrate compared to conventional spray processes such as APS. This results in a much-refined microstructure in the case of the suspension-derived SPS coatings in comparison to the powder-derived APS coatings. The top-view SEM micrograph of the TiC coating shows unmolten spherical particles (see Figure 4a). The Cr_3C_2 coating also showed fewer unmolten particles than those observed in TiC coating (Figure 4b). The reason for lower unmolten particles in Cr_3C_2 (melting point: ~1800 °C [14]) than TiC could be attributed to the significantly lower melting point of Cr_3C_2 compared to TiC (melting point: ~3100 °C). Furthermore, in both the SPS-processed coatings, the top surface morphology did not resemble the cauliflower-like microstructure as reported elsewhere for SPS-processed YSZ coatings, which accompanies the columnar microstructures desired for TBC applications [6].

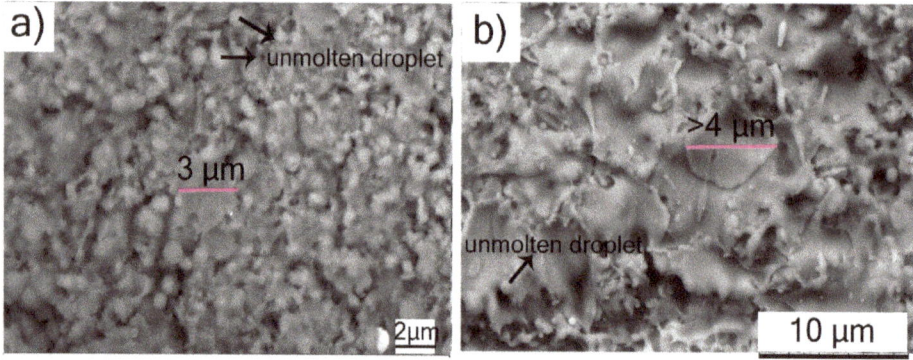

Figure 4. Surface morphology SEM micrographs of SPS-deposited coatings: (**a**) TiC; (**b**) Cr_3C_2.

The low-magnification cross-sectional SEM micrograph of an SPS-processed TiC coating is shown in Figure 5a, and reveals a largely homogeneous microstructure. The corresponding high-magnification cross-sectional SEM micrograph is more revealing, and shows uniformly distributed porosity with very few unmelted particles (Figure 5b). Furthermore, the splat boundaries between successive splats could be discerned, but no inter-pass porosity was obvious, suggesting reasonably good inter-splat cohesion. Additionally, the cross-sectional SEM micrographs did not show any delamination cracks or separation at the splat boundaries, indicating good coating integrity.

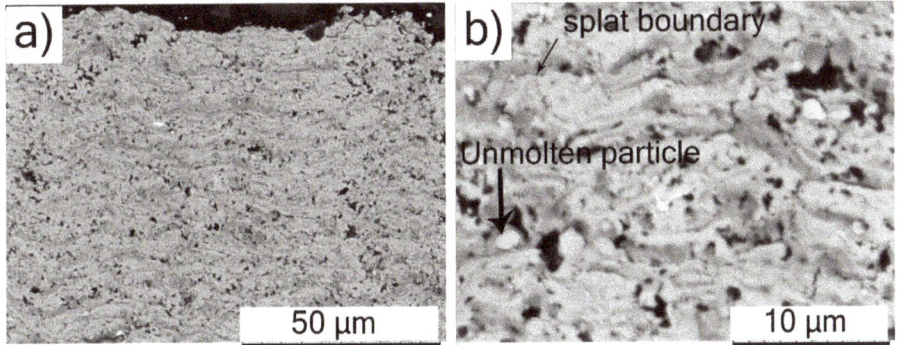

Figure 5. SEM micrographs of cross sections of SPS-deposited TiC coating: (**a**) low magnification; (**b**) high magnification.

The cross-sectional SEM micrograph of the Cr_3C_2 coating at low-magnification in Figure 6a also showed uniform distribution of porosity. The high-magnification cross-sectional SEM micrograph in Figure 6b shows inter-splat boundaries between successive splats. At certain locations, the splat boundaries in the Cr_3C_2 coating were observed to reveal separation between splats, indicating relatively poor cohesion. Notwithstanding the very promising results exhibited by the Cr_3C_2 coating as discussed subsequently, the above microstructural examination suggests that there clearly exists room for further optimization of coating quality by modifying suspension formulation and/or manipulating spray parameters.

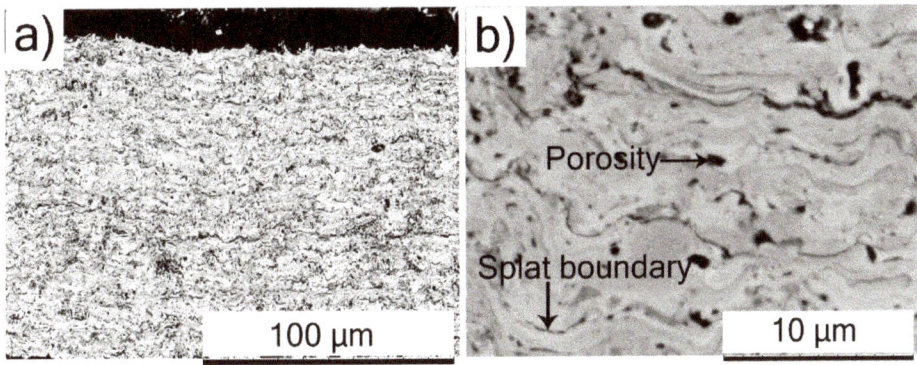

Figure 6. SEM micrographs of cross sections of SPS-deposited Cr_3C_2 coating: (**a**) low magnification; (**b**) high magnification.

Phase analysis of both transition metal carbide coatings by micro-XRD revealed varying extents of oxidation and decarburization occurring in-flight, to show different levels of sub-carbides and oxides in Figure 7a,b. Among various carbide materials, TiC is known for its difficulty in melting and is quite stable over a wide compositional range, between $TiC_{0.97}$ and $TiC_{0.50}$. It is highly challenging to retain the carbides during plasma spraying [15,16]. However, in the present case, the strongest peaks corresponded to titanium oxy-carbide and TiC phases (Figure 7a). This implies minimal decarburization, although considerable amounts of titanium oxides were noted to have formed during the SPS deposition process. The predominant presence of carbide phases can potentially provide better wear resistance by reducing the friction coefficient between the sliding contacts, which was confirmed from the sliding wear results (Table 2). Furthermore, the XRD pattern of the SPS TiC coating in Figure 7a suggests that there could be some amorphous phase formation. Li et al. have reported similar findings related to amorphous phase content in tungsten-carbide-based coatings deposited by HVOF [17].

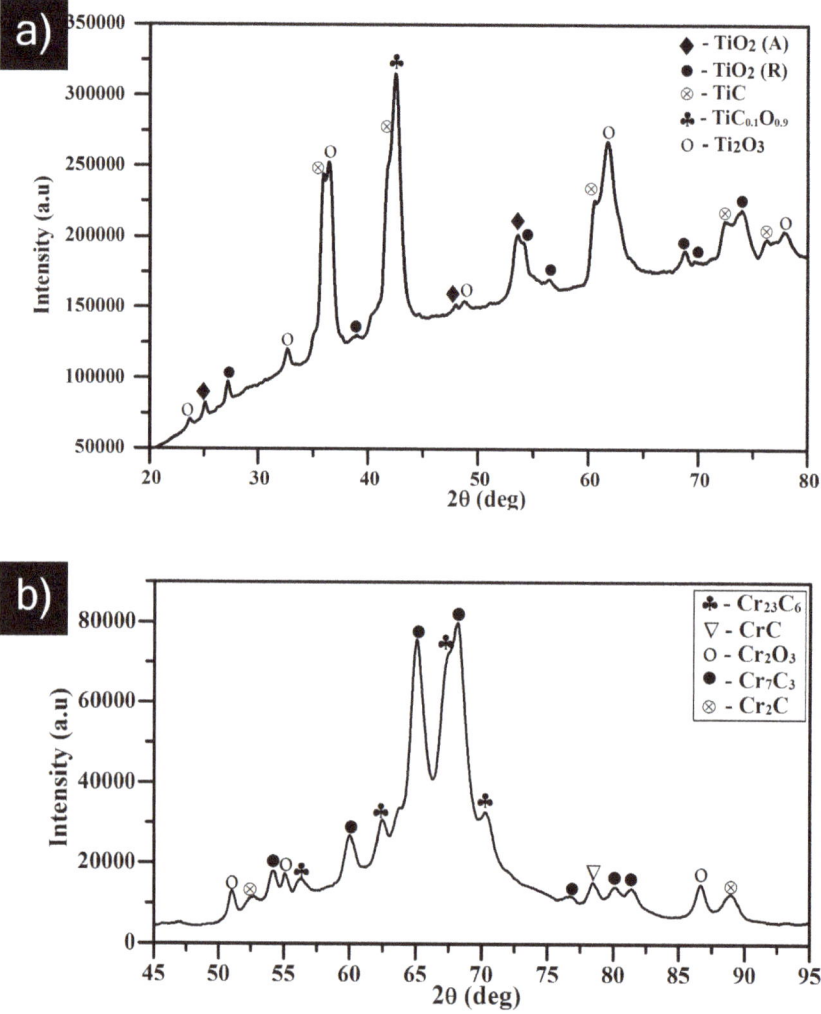

Figure 7. XRD analysis of the as-sprayed surface of (a) TiC coating; (b) chromium carbide coating.

Similarly, plasma/HVOF spraying of Cr_3C_2 typically results in phase transformation into Cr_7C_3 and $Cr_{23}C_6$. It has also been previously reported that Cr_7C_3 forms as a rim over Cr_3C_2 particles [15]. On the other hand, the completely molten chromium carbide (melting point: 1811 °C [14]) preferably precipitates in the form of stable $M_{23}C_6$-type carbides of $Cr_{23}C_6$, which are relatively smaller in grain size [18]. Accordingly, Cr_7C_3 and $Cr_{23}C_6$ phases were detected in case of SPS Cr_3C_2 coatings along with non-stoichiometric CrC phases and oxides of chromium, as shown in Figure 7b.

The TiC coating (12.9 ± 2.2 µm) showed a higher surface roughness (Ra) in as-sprayed condition than the Cr_3C_2 coating (4.6 ± 1.2 µm). It should be mentioned that the TiC feedstock had a relatively lower median particle size (D_{50}) and lower splat size in the as-sprayed condition than did Cr_3C_2. Furthermore, both the coatings were deposited using identical spray parameters. One possible explanation for the observed difference in Ra values could be the difference in the degree of in-flight melting of powder particles because of their vastly different melting temperatures. TiC (melting point: 3100 °C) has a higher melting temperature than Cr_3C_2 (melting point: 1800 °C), which could result in a higher degree of melting in the case of Cr_3C_2 than TiC. As discussed previously, some unmolten TiC particles were also noted in the coating, which could have contributed to higher surface roughness, suggesting need for further feedstock and process optimization. However, it is also pertinent to mention that the surface roughness of the SPS-processed Cr_3C_2 coating in this work was lower than that previously reported for HVOF-processed Cr_3C_2-based coating [19]. For wear application, it is desirable to produce coatings with low surface roughness. Therefore, SPS seems to be a promising processing method to achieve coatings with low surface roughness due to its finer splats compared to APS, HVOF, etc.

The porosity content of both SPS processed carbide coatings was comparable and measured to be in the range of 6–9%, which is higher than ideally desired for wear applications. However, this preliminary attempt at depositing carbide coatings via SPS constitutes a useful basis to minimize porosity content and further improve coating quality through process optimization. The micro-hardness values for SPS-processed Cr_3C_2 (920 ± 70 $HV_{0.1}$) and TiC (980 ± 60 $HV_{0.1}$) coatings were found to be lower than the bulk hardness values for TiC (2850 $HV_{0.1}$) [20] and Cr_3C_2 (1834 $HV_{0.1}$) specimens [21,22]. This could be attributed to the features such as pores and inter-splat boundaries that are typical of thermally sprayed coatings. Regardless of the above, the wear performance of these coatings was suggestive of considerable promise, as discussed below.

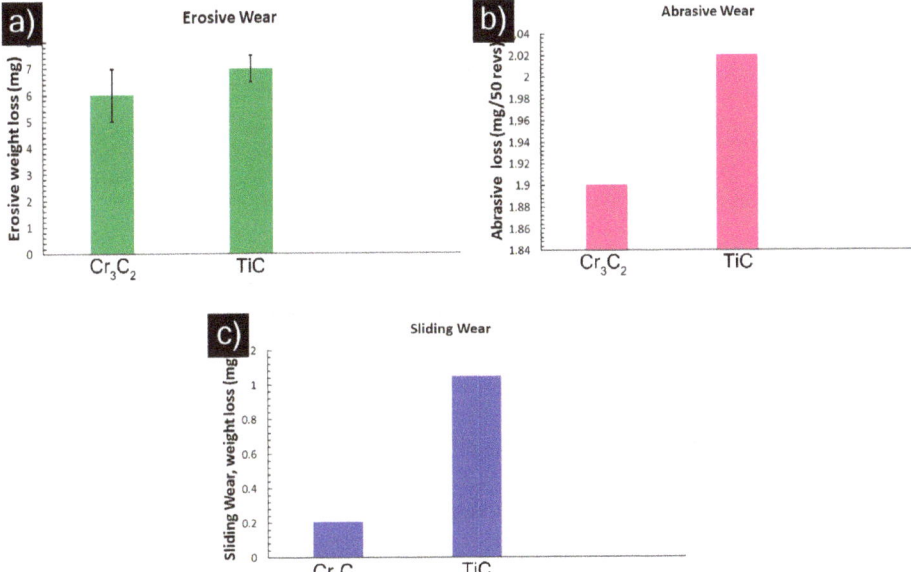

Figure 8. Performance of SPS-processed Cr_3C_2 and TiC coatings under different wear modes: (**a**) erosion; (**b**) abrasion; and (**c**) sliding.

In the wear tests performed under identical test conditions, SPS-processed Cr_3C_2 showed superior erosive and sliding wear resistance to the SPS-processed TiC coating (Figure 8a,c and Table 2). However, the abrasion test results showed comparable performance for the investigated coatings, according to Figure 8b. Note that the hardness for the TiC coating was higher than that of the Cr_3C_2 coating. Furthermore, the cross-sectional SEM micrograph of Cr_3C_2 showed few regions of poor cohesion between the splats. However, its erosion and sliding wear performance was superior to that of TiC. On the other hand, such fine-structured splats in the SPS-processed carbide coatings (in case of both Cr_3C_2 and TiC) could favor improved performance under all wear modes compared to a microstructure with relatively larger splats (APS) for an identical composition. Similar findings related to improved wear resistance compared to conventional coatings was reported by Liang et al. in the case of nanostructured coatings, and the reason was attributed to their enhanced mechanical properties (higher hardness, higher cohesive strength, higher toughness, etc.) [23]. The above results from our preliminary study with carbide suspensions show considerable promise and clearly motivate further, more detailed, investigations. Process optimization and the investigation of different suspension properties (e.g., carbide particle size, solid loading, etc.) and their influence on wear behavior are logical next steps. Furthermore, examining the worn surfaces of the coatings and worn debris (for sliding wear) using SEM/EDS would provide further insights into the wear mechanisms responsible for material removal under different wear modes. These will be performed as a continuation of the present study.

Table 2. Sliding wear results of the investigated coatings.

Sample Identity	Volumetric Wear Loss (mm^3)	Coefficient of Friction (μ)
Cr_3C_2	0.0314	0.60
TiC	0.2129	0.28

4. Conclusions

In this work, we demonstrated for the first time that TiC and Cr_3C_2 coatings could be successfully deposited by suspension plasma spray (SPS) utilizing a feedstock of fine powders. The top-view microstructures of both coatings showed splats that were an order of magnitude lower than those typically observed in APS coatings, indicating the possibility of producing coatings with refined microstructures for wear applications. XRD analysis revealed the presence of the desired carbide phases in the deposited coatings, demonstrating the potential of SPS as a facile route to the deposition of carbide-based coatings. The coatings were shown to have good integrity, and the tribological performance of these coatings was found to be extremely promising under different wear modes (i.e., erosion, sliding contact, and abrasion), motivating their further investigation with and without the addition of binder materials.

Author Contributions: Conceptualization, S.J. and S.B.; methodology, S.B., S.J. and S.M.; formal analysis, K.N., S.M., S.B., S.G.; investigation, S.M., K.N. and S.G.; resources, S.B., S.J., S.G. and N.C.; data curation, S.M. and S.G.; writing—original draft preparation S.M. and S.G.; writing—review and editing, N.C., S.G., S.J.

Funding: This research received no external funding.

Acknowledgments: The authors thank Richard Trache and Christopher Veitsch (Treibacher Industrie AG, Austria) for their assistance in analysis and supply of the carbide suspensions used in this study.

Conflicts of Interest: The authors declare no conflict of interest.

References

1. Hoornaert, T.; Hua, Z.K.; Zhang, J.H. Hard wear-resistant coatings: A review. In *Advanced Tribology*; Luo, J., Meng, Y., Shao, T., Zhao, Q., Eds.; Springer: Berlin/Heidelberg, Germany, 2010; pp. 774–779.
2. Guilemany, J.M.; Nutting, J.; Llorcalsern, N. Microstructural examination of HVOF chromium carbide coatings for high-temperature applications. *J. Therm. Spray Technol.* **1996**, *5*, 483–489. [CrossRef]
3. Fan, W.; Bai, Y. Review of suspension and solution precursor plasma sprayed thermal barrier coatings. *Ceram. Int.* **2016**, *42*, 14299–14312. [CrossRef]

4. Lamuta, C.; Di Girolamo, G.; Pagnotta, L. Microstructural, mechanical and tribological properties of nanostructured YSZ coatings produced with different APS process parameters. *Ceram. Int.* **2015**, *41*, 8904–8914. [CrossRef]
5. Curry, N.; VanEvery, K.; Snyder, T.; Markocsan, N. Thermal conductivity analysis and lifetime testing of suspension plasma-sprayed thermal barrier coatings. *Coat.* **2014**, *4*, 630–650. [CrossRef]
6. Bernard, B.; Queta, A.; Bianchib, L.; Jouliab, A.; Maliéc, A.; Schickd, V.; Rémyd, B.J. Thermal insulation properties of YSZ coatings: Suspension Plasma Spraying (SPS) versus electron beam physical vapor deposition (EB-PVD) and atmospheric plasma spraying (APS). *Surf. Coat. Technol.* **2017**, *318*, 122–128. [CrossRef]
7. Mubarok, F.; Espallargas, N. Suspension plasma spraying of sub-micron silicon carbide composite coatings. *J. Therm. Spray Technol.* **2015**, *24*, 817–825. [CrossRef]
8. Berghaus, J.O.; Marple, B.; Moreau, C. Suspension plasma spraying of nanostructured WC-12Co coatings. *J. Therm. Spray Technol.* **2006**, *15*, 676–681. [CrossRef]
9. Rampon, R.; Marchand, O.; Filiatre, C.; Bertrand, G. Influence of suspension characteristics on coatings microstructure obtained by suspension plasma spraying. *Surf. Coat. Technol.* **2008**, *202*, 4337–4342. [CrossRef]
10. ImageJ. Softonic. Available online: https://imagej.en.softonic.com (accessed on 8 April 2019).
11. ASTM G76-13. *Standard Test Method for Conducting Erosion Tests by Solid Particle Impingement Using Gas Jets*; ASTM International: West Conshohocken, PA, USA, 2013.
12. ASTM G65-16e1. *Standard Test Method for Measuring Abrasion Using the Dry Sand/Rubber Wheel Apparatus*; ASTM International: West Conshohocken, PA, USA, 2016.
13. ASTM G99-17. *Standard Test Method for Wear Testing with a Pin-on-Disk Apparatus*; ASTM International: West Conshohocken, PA, USA, 2017.
14. Matthews, S. Development of high carbide dissolution/low carbon loss Cr_3C_2–NiCr coatings by shrouded plasma spraying. *Surf. Coat. Technol.* **2014**, *258*, 886–900. [CrossRef]
15. Heimann, R.B. *Plasma-Spray Coating: Principles and Applications*; VCH: Weinheim, Germany, 1996.
16. Morozumi, S.; Kikuchi, M.; Kanazawa, S. Plasma spray coating of low Z ceramics on molybdenum. *J. Nucl. Mater.* **1981**, *103*, 279–281. [CrossRef]
17. Li, C.J.; Ohmori, A.; Harada, Y. Formation of an amorphous phase in thermally sprayed WC-Co. *J. Therm. Spray Technol.* **1996**, *5*, 69–73. [CrossRef]
18. Ji, G.C.; Li, C.J.; Wang, Y.Y.; Li, W.Y. Microstructural characterization and abrasive wear performance of HVOF sprayed Cr_3C_2–NiCr coating. *Surf. Coat. Technol.* **2006**, *200*, 6749–6757. [CrossRef]
19. Reyes-Mojena, M.Á.; Sánchez-Orozco, M.; Carvajal-Fals, H.; Sagaró-Zamora, R.; Camello-Lima, C.R. A comparative study on slurry erosion behavior of HVOF sprayed coatings. *Dyna* **2017**, *84*, 239–246. [CrossRef]
20. MEMSnet. Material: Titanium Carbide (TiC), Bulk. Available online: https://www.memsnet.org/material/titaniumcarbideticbulk (accessed on 1 July 2019).
21. MEMSnet. Material: Chromium Carbide (Cr3C2), Bulk. Available online: https://www.memsnet.org/material/chromiumcarbidecr3c2bulk (accessed on 1 July 2019).
22. Hussainova, I.; Jasiuk, I.; Du, X.; Cabassa, D.; Pirso, J. Mechanical properties of chromium carbide based cermets at micro-level. In *Advances in Powder Metallurgy and Particulate Materials—2008, Proceedings of the 2008 World Congress on Powder Metallurgy and Particulate Materials, PowderMet 2008*; Lawcock, R., Lawley, A., McGeehan, P., Eds.; Metal Powder Industries Federation: Princeton, NJ, USA, 2008; pp. 10180–10191.
23. Liang, B.; Zhang, G.; Liao, H.L.; Coddet, C.; Ding, C.X. Structure and tribological performance of nanostructured ZrO_2-3 mol% Y_2O_3 coatings deposited by air plasma spraying. *J. Therm. Spray Technol.* **2010**, *19*, 1163–1170. [CrossRef]

© 2019 by the authors. Licensee MDPI, Basel, Switzerland. This article is an open access article distributed under the terms and conditions of the Creative Commons Attribution (CC BY) license (http://creativecommons.org/licenses/by/4.0/).

Article

High Velocity Suspension Flame Spraying (HVSFS) of Metal Suspensions

Matthias Blum [1], Peter Krieg [1], Andreas Killinger [1,*], Rainer Gadow [1], Jan Luth [2] and Fabian Trenkle [2]

1. Institute for Manufacturing Technologies of Ceramic Components and Composites (IMTCCC), University of Stuttgart, Allmandring 7b, 70569 Stuttgart, Germany; ifkb@ifkb.uni-stuttgart.de (M.B. & P.K. & R.G.)
2. obz innovation GmbH, Elsässer Straße 10, 79189 Bad Krozingen, Germany; info@obz-innovation.de (J.L. & F.T.)
* Correspondence: andreas.killinger@ifkb.uni-stuttgart.de; Tel.: +49-711-685-68320

Received: 6 November 2019; Accepted: 27 January 2020; Published: 30 January 2020

Abstract: Thermal spraying of metal materials is one of the key applications of this technology in industry for over a hundred years. The variety of metal-based feedstocks (powders and wires) used for thermal spray is incredibly large and utilization covers abrasion and corrosion protection, as well as tribological and electrical applications. Spraying metals using suspension- or precursor-based thermal spray methods is a relatively new and unusual approach. This publication deals with three metal types, a NiCr 80/20, copper (Cu), and silver (Ag), sprayed as fine-grained powders dispersed in aqueous solvent. Suspensions were sprayed by means of high-velocity suspension spraying (HVSFS) employing a modified TopGun system. The aim was to prepare thin and dense metal coatings (10–70 µm) and to evaluate the process limits regarding the oxygen content of the coatings. In case of Cu and Ag, possible applications demand high purity with low oxidation of the coating to achieve for instance a high electrical conductivity or catalytic activity. For NiCr however, it was found that coatings with a fine dispersion of oxides can be usable for applications where a tunable resistivity is in demand. The paper describes the suspension preparation and presents results of spray experiments performed on metal substrates. Results are evaluated with respect to the phase composition and the achieved coating morphology. It turns out that the oxidation content and spray efficiency is strongly controlled by the oxygen fuel ratio and spray distance.

Keywords: high-velocity suspension flame spraying; copper; silver; NiCr 80/20; metal coatings

1. Introduction

In recent years the high-velocity suspension flame spraying (HVSFS) process, an evolution of the well-established HVOF process that allows for spraying of liquid feedstock instead of finely grained spray powders, has been developed. Whereas the majority of studies focus on the spraying of oxide ceramics, such as aluminum oxide [1,2], titanium oxide [3,4], or zirconium oxide [5], glasses [6], and biomaterials [7,8], there are not many studies on the topic of suspension sprayed metal coatings. Among these, more focus can be found on the topic of suspension plasma spraying [9,10], while only few publications cover the suspension flame spraying of metals, for instance Inconel alloys [11].

The use of suspensions in the flame spraying process allows for the processing of submicron- and nanopowders to form finely structured coatings with reduced coating thickness, which can lead to an improvement of the mechanical, thermal and chemical properties of the coatings [12]. While suspension flame spraying offers many advantages, it also increases the complexity of the process due to the additional parameters (such as suspension formulation) that have to be considered. In the case of metals, the suspension formulation is especially important and challenging due to the surface reactivity and high density of the involved metal powders.

In this study, suspensions from three metals were prepared and sprayed via HVSFS: NiCr, silver, and copper. While powder and wire spraying of NiCr alloys and Cu is more common in thermal spraying, Ag has not been in the focus; and to our knowledge, none of these materials have been applied in suspension spray processes so far. The aim of this study is to create metal coatings with a reduced thickness in the range of 10–50 µm. The amount of oxides in the coating should be as low as possible, especially for Cu and Ag. For Ag, a high deposition efficiency is also mandatory to keep material costs in a reasonable range.

NiCr coatings are mostly used as protective coatings or for heating elements and can be sprayed by various thermal spray techniques [13,14]. While copper can also be deposited by different thermal spray techniques, cold gas spraying of copper offers low oxidation and porosity, which results in good electrical properties of the coatings [15]. Silver is a less common material in thermal spraying. There are only very few studies on the cold gas spraying of silver or silver composites for electrical contacts [16,17]. Silver is also well known for its antibacterial effect. Thin suspension-sprayed coating may also serve as an antibacterial layer in medical applications.

Ultrafine Ag powder can also be incorporated as a dopant in biomedical coatings to display their antibacterial properties. This has been examined in recent studies for different calcium phosphate-based coatings. One route uses HVSFS with mixed Ag/HAp, TCP, or bioglass containing suspensions [18], the second route uses a standard plasma spraying process with agglomerated powders containing Ag particles [19].

As already mentioned, one of the main competing spray method for spraying soft metals like copper or silver is cold gas spraying (CGS). A comparative study with CGS has been recently published by some of the authors [20]. Copper and silver powders were sprayed using a Kinetiks 4000 system (CGT Cold Gas Technology GmbH, Ampfing, Germany). As powder feeder operated spray systems are restricted to a minimum powder grain size of >5 µm (typically grain size distribution in the range of 5–25 µm) coatings show higher roughness and thickness than suspension sprayed coatings, but it is well known that CGS produces a more or less pore-free coating structure when well-plasticizing metals, like aluminum, copper, or silver, are used as a feed stock. Due to the coarser spray particles, CGS coatings are usually limited to a minimum thickness of approx. 20 µm, but are theoretically not limited in regard to a maximum thickness value. In contrast, HVSFS can go below 10 µm, but is limited at higher thickness due to the increasing formation of residual stresses in the coating when increasing the coating thicknesses. As CGS is operated on a significantly lower thermal level of approx. 800 °C, the amount of oxides in the coating can be usually kept significantly lower than in HVSFS coatings. This is especially relevant for electrical applications, where a high electrical conductivity is demanded. Therefore, spraying copper is one of the main industrial application in CGS industry. In any case, a more elaborate comparative study is necessary to compare electrophysical properties of Cu and Ag coatings sprayed with the two methods.

2. Materials and Methods

2.1. Preparation and Characterization of the Metal Suspensions

Three different metal powders were selected for this study. Available manufacturer data have been summarized in Table 1. The first was a gas atomized NiCr powder (H.C. Starck, Goslar, Germany), the second was a chemically derived copper powder (Evochem GmbH, Offenbach, Germany) and the third was an agglomerated silver powder (Metalor Technologies SA, Neuchatel, Switzerland). As can be seen in Figure 1, the NiCr powder shows a perfect spherical structure, which is generally achieved in the gas atomization process. The agglomerated and sintered silver powder also shows a more or less spherical structure, whereas the copper particles show a dendritic and irregular structure, resulting from the chemical precipitation process.

Table 1. Overview of the commercial powders used for suspension preparation. All data in this table are according to supplier datasheets.

	NiCr 80/20	Cu	Ag
Manufacturer	H.C. Starck	Evochem GmbH	Metalor Technologies SA
Notation	Amperit 251.051	14/200	P747-35
Density [g/cm^3]	8.31	8.93	10.49
D10 [µm]	4	-	1.8
D50 [µm]	7.8	-	6.2
D90 [µm]	13.6	-	20
Powder type	Gas atomized	Chemically derived	Agglomerated and sintered
Morphology	spherical and dense	irregular dendritic	spherical

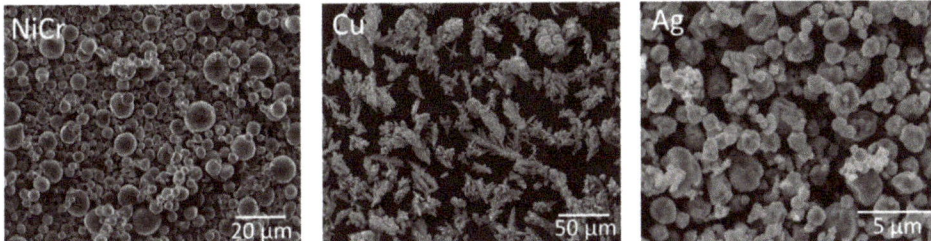

Figure 1. SEM images of the metal powders used in this study.

The materials were dispersed in deionized water with a solid content of 18 wt% for the copper suspension and 30 wt% for the NiCr and the silver suspension. In contrast to organic solvents like isopropanol or ethanol, deionized water does not deliver combustion enthalpy and the evaporation enthalpy is significantly higher (water: 2.26 kJ/g; isopropanol: 0.66 kJ/g; ethanol: 0.84 kJ/g). Therefor it was used as dispersant to reduce the total combustion enthalpy, leading to a reduction in flame temperature and thus reducing oxidation of the metal particles. To improve the stability and to prevent sedimentation and clogging during the process, suspensions were stabilized using an organic dispersant, an anti-foaming agent and a rheology additive. The amount of solid content has been chosen to adjust the viscosity behavior of the suspensions. In case of copper, the solid content could not be increased to higher value due to the irregular particle morphology that promotes clogging in the feedline. It has been described in more detail in a previous publication [21]. The particle size distribution was measured by laser diffraction (Mastersizer 3000, Malvern Instruments, Malvern, UK). The rheological properties of the suspensions were characterized using a MCR 302 rheometer (Anton Paar GmbH, Graz, Austria). The measurements were performed at 25 °C using a double-gap cylinder.

2.2. Coating Deposition and Characterization

For the HVSFS process, a modified high-velocity flame spray torch type TopGun (GTV GmbH, Luckenbach, Germany) was employed. In the modified torch, the powder injector is replaced with a suspension injector unit, which allows for injection of the suspension axially into the combustion chamber. The suspension injector is equipped with a simple turbulence nozzle having an orifice diameter of 0.5 mm. A more detailed description of the HVSFS process can be found in [12].

The coatings were deposited on planar stainless steel (X6CrNiMoTi17-12-2) substrates. In case of some Ag-coatings pure copper samples were alternatively used as substrate. Dimensions of all substrate samples were 50 × 50 × 3 mm. Surface activation was done by grit-blasting using F120 corundum at a pressure of 5 bar. Prior to coating operation, all specimens were cleaned with acetone. Whereas the TopGun-system allows for the use of a wide variety of fuel gases, all coatings in this study were sprayed using an ethene-oxygen mixture. The torch was operated either at a stoichiometric ethene-oxygen ratio (denoted as $\lambda = 1$) and, for comparison also at different sub-stoichiometric ratios, thus having a surplus of ethene (denoted as λ = 0.57; 0.63, and 0.87, respectively). The idea in mind is to have a reducing flame chemistry that may reduce oxidation of metal particles in the spray process.

The torch was operated on a six-axis robot describing a meander movement in front of the planar substrates with a torch speed of 500 mm/s. Two air cooling nozzles were mounted on the torch to simultaneously cool the substrate during spray process using compressed air. The two cooling spots are located on both sides near the spray focus on the substrate surface. Samples were sprayed at five distances: 60 mm, 75 mm, 90 mm, 105 mm, and 120 mm and a suspension feed rate of 40–55 g/min. An overview of the spraying parameters can be found in Table 2. It should be noted that for all samples a fixed amount of torch passes was used (NiCr and Cu: eight passes; Ag: four passes) also holding constant the suspension feed rate for each system. This allows for comparison of relative deposition efficiencies assuming comparable porosity values for all coatings. In case of HVSFS coating porosity typically is in the range of 0–5%. Thus, when calculating DE from coating thickness the error is in the same magnitude (0–5%). In the case of silver, the deposition efficiency was measured following DIN EN ISO 17836 according to Equation (1):

$$DE\,[\%] = \frac{m_c}{m_t} \times 100 \tag{1}$$

m_c: mass of coating
m_t: theoretically sprayed mass

Table 2. Overview of the HVSFS spraying parameters discussed in this paper.

Parameter	NiCr		Cu		Ag	
Normalized ethene-oxygen ratio λ	0.87	1	0.57/0.63	1	0.63	1
*TGFR (oxygen + ethene) (slpm)	180	200	190/230	230	230	160
**GFR oxygen (slpm)	130	150	120/150	58	150	120
**GFR ethene (slpm)	50	50	70/80	172	80	40
Spray distance variations (mm)	For all three materials: 60; 75; 90; 105; 120					
Meander offset (mm)	For all three materials: 3					
Torch speed (mm/s)	For all three materials: 500					
Length of combustion chamber (mm)	22	22	22	22	12	12
Suspension feed rate (g/min)	55		43		52	
Number of torch passes	8	8	8	8	4	4

*TGFR = Total gas flow rate; **GFR = Gas flow rate; slpm = standard liters per minute.

Further spray parameters that are listed in Table 2 are: ethene-oxygen ratio λ, total gas flow rates (TGFR), separate gas flow rates (GFR) for ethene and oxygen. Spray distance variations, meander offset and torch speed have identical values for all three materials.

Cross-section images of the coatings were taken using a Leica MEF4M (Leica GmbH, Wetzlar, Germany). The coating thickness was measured on eight different points of these cross-sections.

The surface structure of the coatings was investigated using a white light interferometry (Bruker ContourGT-K, Bruker, Mannheim, Germany) using a 5× objective. Three surface areas with a size of 1.2 × 0.9 mm were evaluated.

A comparative microhardness measurement study was performed to analyze the coating properties for the different spray parameters, i.e., spray distance, λ, and total gas volume. Vickers microhardness ($HV_{0.1}$) measurements were carried out on polished cross-sections using a Fischerscope H100 instrument (Helmut Fischer, Sindelfingen, Germany). A set of 20 measurements were performed on one probe and evaluated statistically. The study was made in accordance to DIN EN ISO 12577.

High-resolution SEM pictures and the EDX measurements were performed using a SEM DSM 982 Gemini (Zeiss AG, Oberkochen, Germany).

XRD analysis on coatings have been carried out using a X´Pert MPD diffractometer using Cu Kα (40 kV/40 mA) in combination with a Panalytical X´Celerator detector (both from Panalytical

GmbH, Herrenberg, Germany). The used database was taken from International Centre for Diffraction Data (JCPDS ICDD). For qualitative analysis "Crystallographica Search Match" software (Oxford Cryosystems Ltd., Long Hanborough, UK) has been used. The Siroquant software (Sietronics Pty Ltd., Mitchell, ACT 2911, Australia) was used for quantitative Rietveld analysis.

3. Results and Discussion

3.1. Suspension Characterization

Suspensions were evaluated regarding their particle size distribution, their sedimentation behavior and their rheological properties. It was found, that particles tend to form agglomerates in the suspension, which can be redispersed using ultrasonic treatment. In Figure 2, the particle size distribution of the stabilized silver suspension is displayed. If the sample is treated ultrasonically, the measured particle size is reduced significantly. However, treatment of the suspension using ultrasonic has to be done carefully, as it degrades the stabilization of the suspension, which leads to a faster sedimentation of the suspended particles.

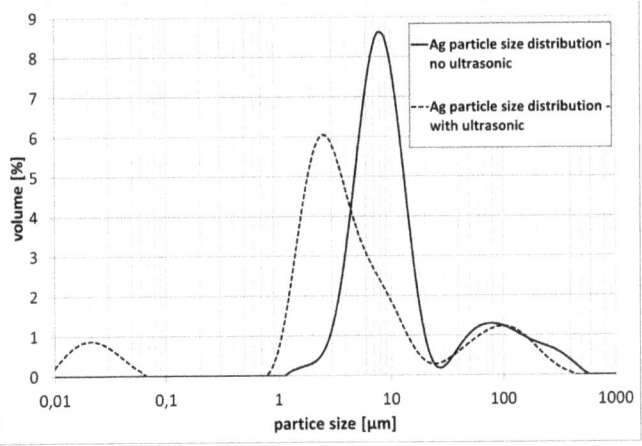

Figure 2. Particle size distribution of silver suspension—with and without ultrasonic treatment.

By applying ultrasonic treatment, average particle size d50 in silver suspension drops from 9.5 µm to 3.6 µm. This effect can also be observed for the copper suspension, showing a measured average particle size d50 of 64.8 µm without, and 34.6 µm with, ultrasonic treatment. Especially, agglomerated powders show a weak mechanical cohesion and easily form fragments. This effect is even stronger when applying ultrasonic treatment. The NiCr suspension does not show this effect (d50 = 10.8 µm). This can be mainly explained by the powder particle properties. The gas atomized particles are mechanical stable and do not tend to break.

The flow curves of the suspensions are shown in Figure 3. All three suspensions show shear thinning behavior. The observed difference between the NiCr suspension and the Ag and Cu suspensions is probably due to a slightly lower concentration of rheology additive in the NiCr suspension. Particle morphology also has an effect on rheological behavior of the suspension.

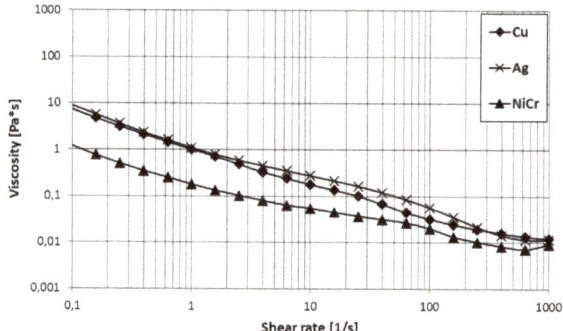

Figure 3. Flow curves of the Cu, Ag, and NiCr suspensions used in this study.

3.2. Coating Structure and Properties

3.2.1. NiCr Coatings

In Figure 4, a collection of cross-sections of HVSFS sprayed NiCr coatings at different spraying distances and oxygen-to-ethene ratios are shown, the bright areas correspond to metal-rich, dark areas to oxide phases.

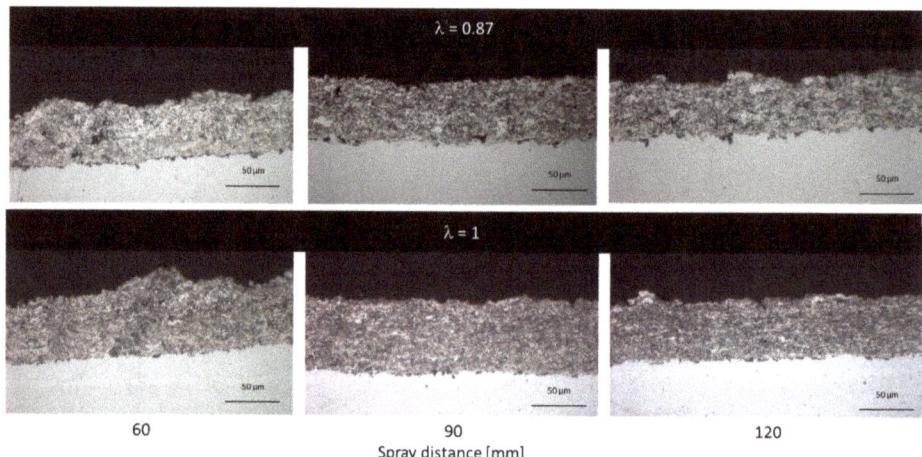

Figure 4. Optical microscope cross-section images of NiCr coatings sprayed with two different λ (0.87 and 1) and with three different spray distances (60, 90, and 120 mm).

At a very close spraying distance of 60 mm, an increased formation of defects can be observed. The formation of these defects begins near the substrate and increases in size during the build-up of the coating, resulting in a cone-shaped defect. Fauchais et al. described the formation and possible reasons of this defects [22–24]. These defects grow at an angle from the surface and show an increase in porosity (Figure 5). Consequently, the surface roughness Sa (shown in Figure 6) of these coatings is significantly higher. This effect is more pronounced at a stoichiometric oxygen-to-ethene ratio.

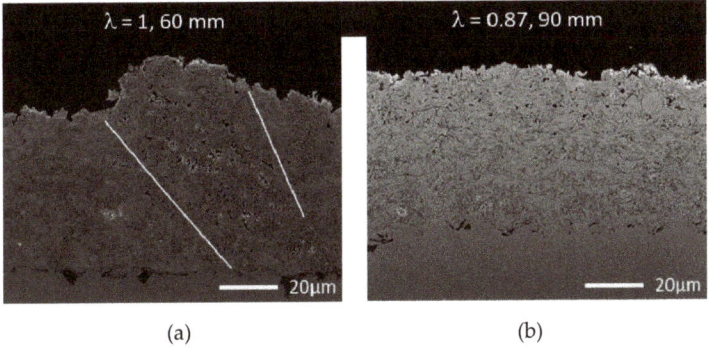

Figure 5. SEM image of NiCr coatings. (**a**) A typical cone-shaped defect found in NiCr coatings at 60 mm spray distances ($\lambda = 1$). (**b**) Spray parametres: $\lambda = 0.87$ (<1); spray distance = 90 mm. Taken from [21].

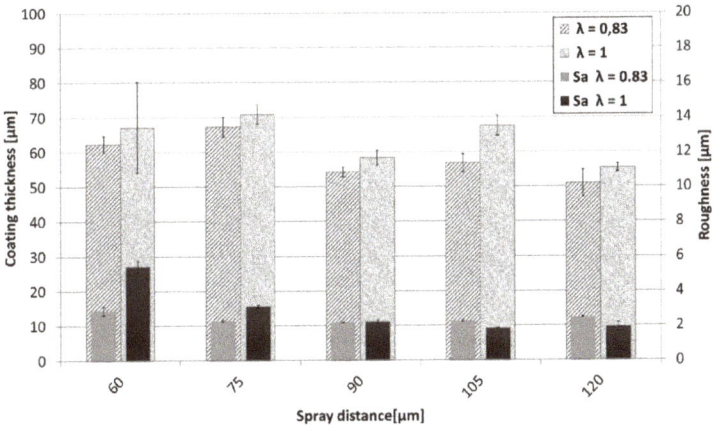

Figure 6. Comparison of coating thickness and roughness values of the NiCr coatings sprayed at different spray distances and λ values.

The variation of the oxygen-to-ethene ratio λ clearly affects the oxidation of the NiCr particles. The coatings sprayed at a surplus of ethene ($\lambda < 1$) contain more metallic particles (visible as the bright phase in the light micrograph), while the coatings sprayed at stoichiometric parameters show higher oxidation (darker phases visible in the micrographs in Figure 4). A similar approach is described in detail by Förg et al. in the case of suspension spraying of Cr_3C_2. Varying λ allows for an adjustment of the formation of Cr_2O_3 [25]. Consequently, for NiCr coatings, it is possible to adjust the electrical conductivity and resistance as demanded by the desired application (i.e., heating elements) [26].

As all coatings have been prepared using the same amount of spray passes (= 8), the achieved coating thickness can serve as an indicator for deposition efficiency. These are summarized in Figure 6 together with the Sa values. Deposition efficiency slightly rises with decreasing spray distance. When spraying under stoichiometric conditions ($\lambda = 1$), slightly higher deposition efficiencies can be observed. As already mentioned, at lower spray distances more coating defects occur that contribute to a significantly increased surface roughness.

The microhardness measurement results are summarized in Figure 7. Hardness values differ significantly and show a dependency of spray distance, λ and total gas volume. For lower gas volumes (180 and 200 L/min), the spray distance clearly controls hardness values. As observed in the microscope images higher spray distances show higher porosities and correspondingly lower hardness values. On the other hand, a higher λ together with a slightly higher total gas volume increases oxidation significantly. Doubling the total gas volume doubles the hardness value. Both factors may contribute: the higher kinetic impact leads to a denser coating structure thus increasing the hardness; the significantly increased heat impact on the other hand leads to a strong increase in oxide species which in turn will increase hardness.

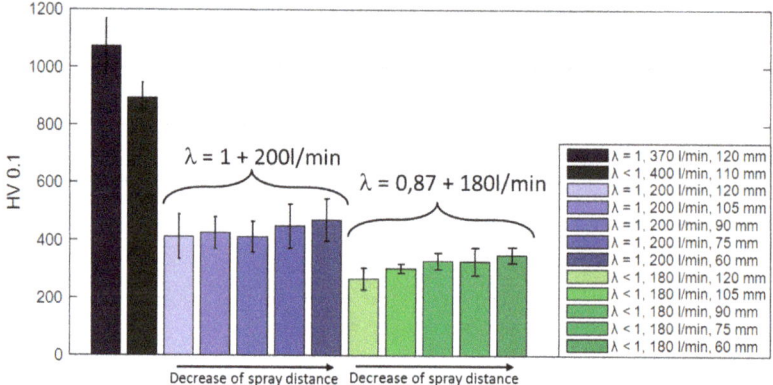

Figure 7. Comparison of Vickers hardness HV0.1 for NiCr coatings sprayed with different λ (0.87 and 1) and at different spray distances (60; 75; 105; 110; 120 mm).

XRD analysis has been carried out to analyze type and quantity of the oxide phases. The NiCr coating sprayed with λ = 0.87 (also noted as λ < 1 in Figure 7) and a spray distance of 90 mm was analyzed for this purpose, please refer to Figure 8. Based on the Rietveld method, a quantitative analysis was performed to learn more about the oxidation species that form during spraying. Achieved results are summarized in Table 3. Additionally, to the metal phases, two major types of oxides can be found in the coating: Spinel (Cr_2NiO_4) and bunsenite (NiO). From this data it can be concluded that the oxide fraction in the coating is in the range of 25%–30%.

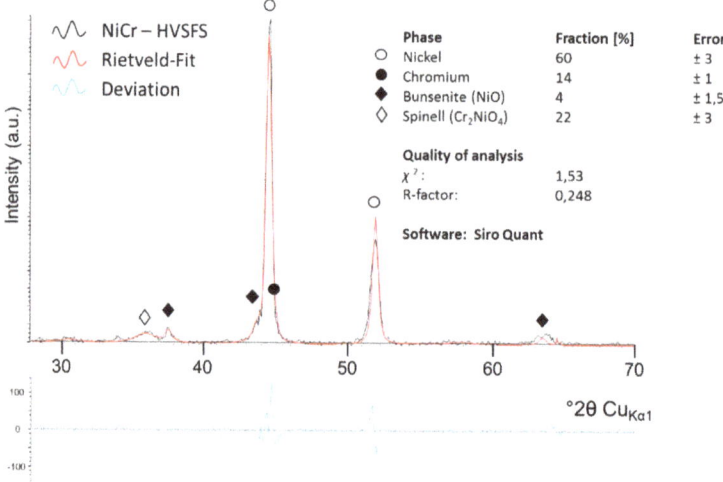

Figure 8. Representative XRD spectrum of a HVSFS sprayed NiCr coating. A Rietveld fit has been applied to the spectrum. Spray parameters (λ = 0.87; total gas flow: 180 L/min; spray distance: 90 mm).

Table 3. Rietveld analysis taken from XRD spectrum in Figure 8.

Phase (Ref. No.)	Fraction (wt%)	Error (wt%)
Nickel (000-04-0850)	60	±3
Chrome (000-01-1250)	14	±1
Spinell Cr_2NiO_4 (000-47-1049)	22	±3
Bunsenit NiO (000-75-0198)	4	±1,5

A comparison of the oxide content for two different total gas flows (180 vs. 370 L/min) shows a significant increase of the oxide peaks in XRD. An increase of the total gas flow from 180 to 370 L/min leads to a doubling of the heat flux as well as a strong increase of the turbulence on the surface. This leads to a higher substrate temperature and thus to higher oxidation rates. Additionally, the slightly higher spraying distance could have an influence. In Figure 9, both spectra are shown for comparison. The relevant oxide peaks are significantly increased. A quantitative Rietveld evaluation estimates the amount of oxide above 50%.

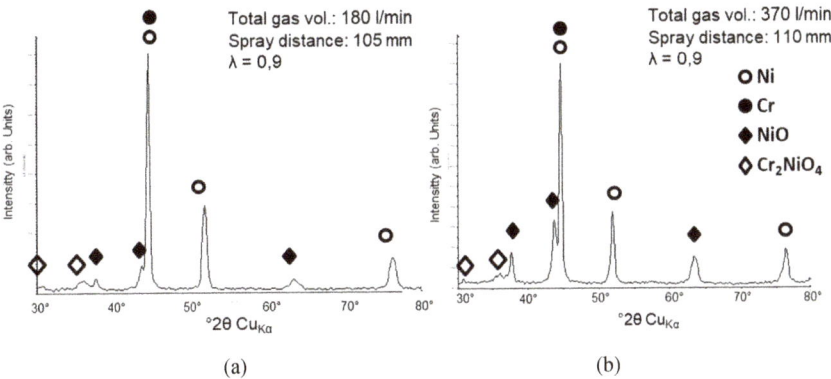

Figure 9. Comparison of XRD spectra for NiCr coatings sprayed with two different total gas flow amounts (180 vs. 370 L/min) as noted in the graph. λ was kept constant, and the spray distance differs slightly. ◆ symbols highlight the most prominent oxide peaks that were identified according to Figure 8.

3.2.2. Cu Coatings

The cross-sections of the copper coatings, as shown in Figure 10, show the difference in oxidation between the stoichiometric and the sub-stoichiometric parameters. At λ < 1, there are distinguishable metallic and oxide layers, which are more pronounced at closer spraying distances. This is presumable due to the higher heat flux at close distances, resulting in a post-oxidation process of the surface layer of the coating.

In contrast, the coatings sprayed at stoichiometric ratios show a more homogeneous distribution of oxide phases. While the increased heat flux at close distance will also influence the oxide formation, the dominating mechanism here appears to be the in-flight oxidation of the particles during the spraying process. The cross-sections show dense microstructures for all copper coatings with low porosity, suggesting that a sufficient melting of the particles during the spraying process has occurred.

As already discussed in the previous section, the achieved coating thickness again can serve as an indicator for deposition efficiency. Coating thicknesses and surface roughness of the Cu coatings, as shown in Figure 11, indicate slightly higher deposition efficiencies for the stoichiometric parameters as well as higher values for the surface roughness, which decreases with increasing spraying distance. Interestingly, when directly compared to NiCr, the spray distance seems to have a minor influence on these values.

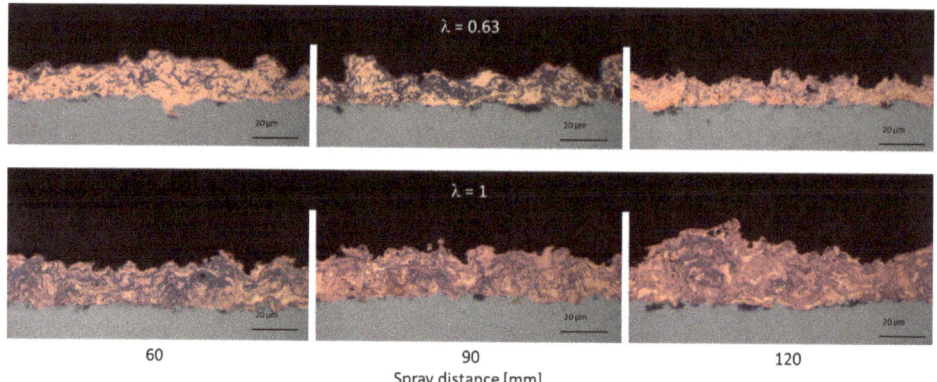

Figure 10. Optical microscope cross-section images of the Cu coatings sprayed with two different λ (0.63 and 1) and with three different spray distances (60, 90, and 120 mm). Taken from [21].

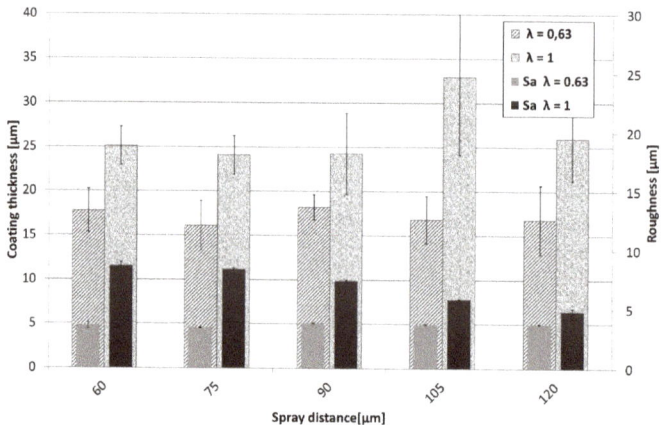

Figure 11. Comparison of coating thickness and roughness values of the Cu coatings sprayed at different spray distances and λ values.

For comparison a set of samples were sprayed with a reduced λ of 0.57, because it was observed that oxidation can be further reduced. Interestingly, oxidation now occurs more in the outer part of the coating, and is somewhat lower near the interface. Respective coatings are displayed in Figure 12.

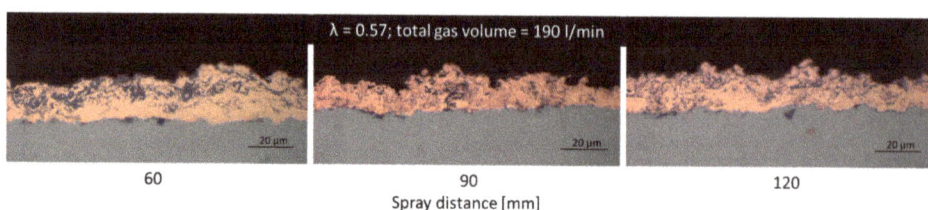

Figure 12. Optical microscope cross-section images of Cu coatings sprayed with a reduced of λ = 0.57 with increasing spray distance (from left to right). Applied spray parameters: λ = 0.57; total gas volume = 190 L/min.

As oxidation is clearly visible in all samples, EDX analysis was performed to analyze the distinguishable phases in regard of their Cu and O content for to Cu coatings at different λ (0.57 and 1) and two different total gas volumes (190 and 230 slpm); refer to Figure 13. For both Cu coatings, a higher amount of oxygen can be detected in the dark phase (6–7 wt%). The bright phase is more pronounced in sample A (λ = 0.57; total gas volume = 190 slpm) with low oxygen content (0.5–1 wt%), whereas in sample B (λ = 1; total gas volume = 230 slpm) metal and oxide lamellas are more densely mixed and cannot be clearly separated by EDX. Thus, the oxygen content of the brighter phase is measured at a higher level (between 2–3 wt%). However, the polished surface contains a high amount of impurities, which is due to the somewhat difficult conditions when preparing cross-sections of the smooth copper material in metallography (noted as other in Table 4). There is a high risk of working in impurities into the polished surface during the polishing process.

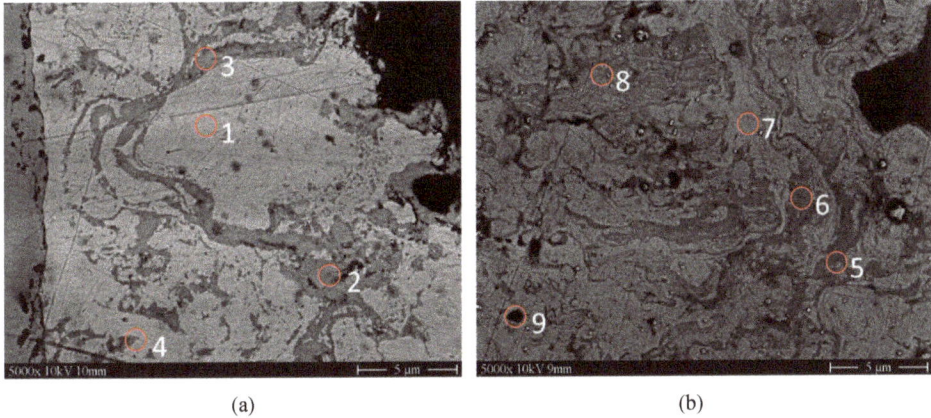

Figure 13. SEM images of two Cu-coatings prepared for EDX-analysis. Left: (**a**) λ = 0.57; total gas volume = 190 slpm; Right: (**b**) λ = 1; total gas volume = 230 slpm. Taken from [21].

Table 4. EDX-Cu and O wt% values achieved from EDX on the samples shown in Figure 13.

Ref. No (Refer to Figure 13)	Cu (wt%)	O (wt%)	Other (wt%)
1	95.32 ± 0.6	0.71 ± 0.13	3.97 ± 0.31
2	87.79 ± 0.45	7.46 ± 0.10	5.74 ± 0.19
3	95.76 ± 0.56	0.69 ± 0.07	3.55 ± 0.27
4	92.66 ± 0.44	0.85 ± 0.04	6.49 ± 0.42
5	93.09 ± 0.47	6.91 ± 0.11	-
6	95.11 ± 0.47	4.63 ± 0.1	0.25 ± 0.18
7	97.09 ± 0.45	2.64 ± 0.09	0.27 ± 0.04
8	95.11 ± 0.48	4.89 ± 0.10	-
9	97.06 ± 0.48	2.94 ± 0.09	-

XRD analysis of HVSFS sprayed copper shows clearly distinguishable peaks for metallic Cu, and oxide species namely copper(II)oxide (cupric oxide, CuO) and copper (I) oxide (cuprous oxide, Cu_2O). Figure 14 shows the spectrum for a coating with spray parameters identical to those shown in Figure 12, left (λ = 0.57; spray distance = 60 mm and total gas volume = 190 l/min).

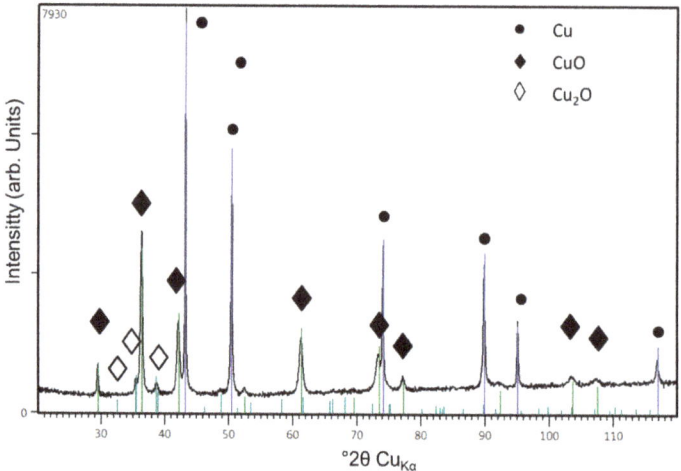

Figure 14. XRD spectrum of HVSFS sprayed Cu coating shown in Figure 12 on the left ($\lambda = 0.57$; spray distance = 60 mm and total gas volume = 190 l/min). CuO and Cu_2O. Strongest peaks are labeled.

3.2.3. Ag Coatings

When spraying silver coatings, it was observed, that in order to achieve a suitable coating, the appropriate parameter window appears to be much narrower than for NiCr and copper. In a parameter pre-study, coatings have been sprayed on mild steel and on copper substrates as can be seen in Figure 15. The colors of the coatings after spraying differ significantly depending on what spray distance and what λ has been chosen. Colors change from brown, yellow to a bright white. This behavior is probably due to formation of different amount of oxide species in the coatings.

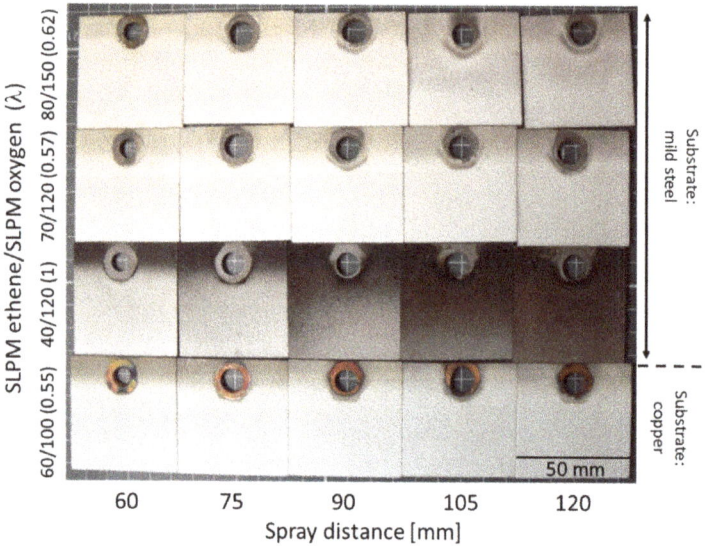

Figure 15. Color scheme of silver coating surfaces sprayed at different parameter sets worked out in a pre-study. Size of each sample is 50 × 50 mm.

Respective cross-sections of the coatings for spray distances at 60, 90, and 120 mm are shown in Figure 16. The stoichiometric parameters yield a lower coating thickness and, consequently, a lower deposition efficiency at all spraying distances. Due to the higher spraying distance, the residual time of the particles in the flame is increased and the formation of oxide species may be favored. As a consequence, the coating shows a more and more greyish color at higher spray distances compared to the bright color at closer ones.

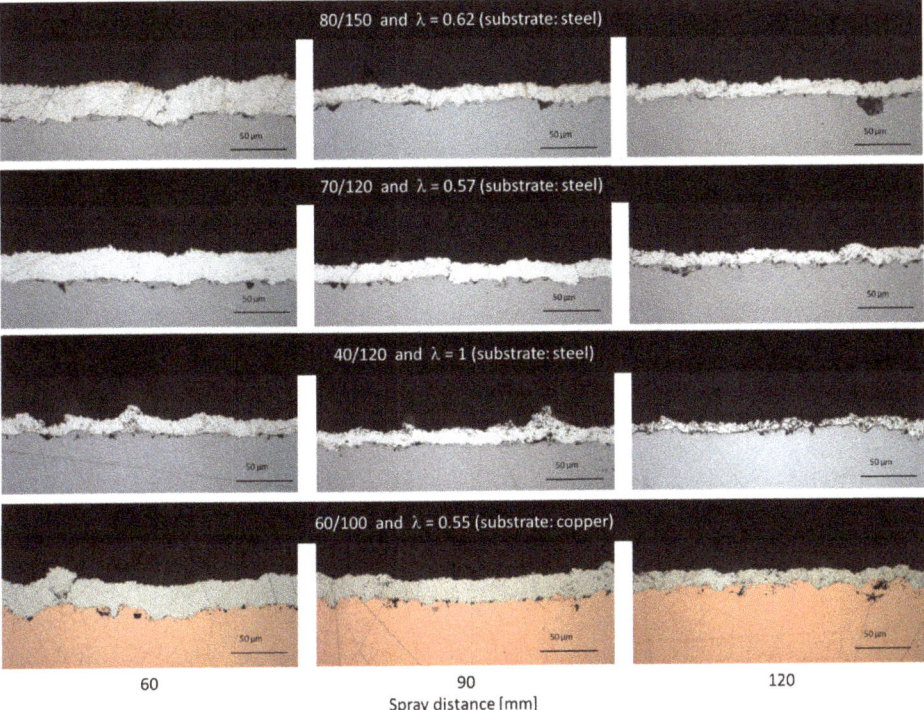

Figure 16. Optical microscope cross-section images of the Ag coatings on mild steel and Cu substrate from samples in Figure 15 chosen for 60, 90, and 120 mm spray distances. Please note that in respect of gas ratios and substrate type, images are organized in the same systematic as in Figure 15.

The cross-sections, as well as the color of the coatings, suggest that best results are achieved at sub-stoichiometric ratios with a distinct surplus of ethene. Although the flame temperature at this ratio should be somewhat higher compared to $\lambda = 1$, there is less formation of oxides. Apparently, the process needs a certain degree of reducing effect to overcome oxidative reactions promoted by the gas mixture itself and the surrounding air. Figure 17 shows the Ag coating at a higher resolution in SEM. It corresponds to the coating shown in Figure 16 in the fourth row, sprayed with 90 mm spray distance. However, in SEM there is no evidence for the presence of oxides. Figure 17b reveals the fine micropore structure that can be found in the coatings.

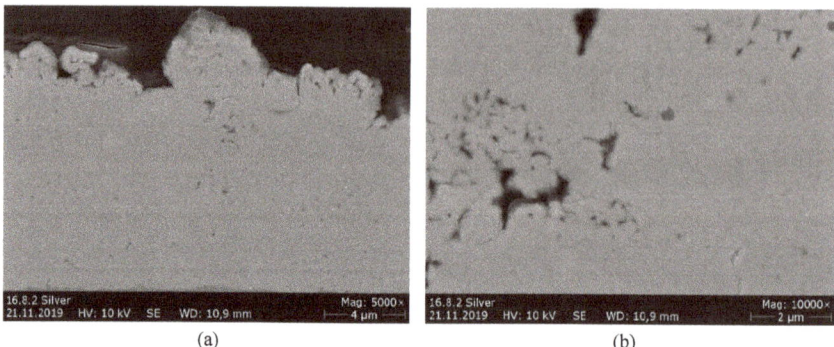

Figure 17. SEM images of Ag-coating on copper substrate with parameter set 60/100, λ = 0, 55, 90 mm, (refer to Figure 16 last row). (**a**) near surface (**b**) details of pore structure.

The coatings sprayed at closer distances and with a sub-stoichiometric oxygen-ethene-ratio are dense and no inter-layer porosity is visible. In contrast to the NiCr and copper coatings, the silver coatings show a pronounced correlation at sub-stoichiometric parameters between spraying distance and coating thickness.

Considering the high costs of silver powder feedstocks, deposition of thin coatings with high efficiency is an interesting approach. With a coating thickness of up to 30 µm deposited in four torch passes, the HVSFS process allows for a further reduction of the coating thickness. With 60 mm spray distance, it seems possible to perform 10 µm thick coatings in one torch pass. However, the short spraying distances result in an increased thermal load of the sample. Depending on the substrate and the desired coating thickness, a more elaborate cooling concept is needed.

Comparative XRD spectra are shown in Figure 18 for coatings sprayed with a gas flow rate of 70/120 and varying spray distances (60–120 mm; refer to Figure 15, second row). Only the peaks of pure silver are detectable, with peak intensity increasing towards lower spray distances. It was not possible to certainly identify oxide species in the XRD spectra.

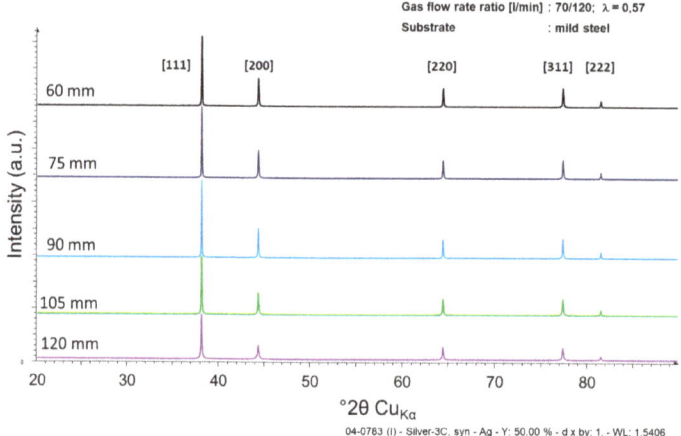

Figure 18. XRD spectrum of HVSFS sprayed Ag coatings as a function of spray distance (assigned as 60, 75, 90, 105, and 120 mm in the graph). Data correspond to samples in Figure 15, second row (60, 90, 120 mm).

Achieved coating thickness and coating surface roughness show a clear dependency of spray distance and the ethene-oxygen ratio λ, as can be seen in Figure 19. Best coating structure and highest coating thickness was reached with sub-stoichiometric values (λ in the range of 0.56–0.63). The achieved coating thickness was in the range of 15–30 µm, sprayed with four torch passes, refer to Figure 19. Shorter spraying distances lead to higher deposition rates and accordingly demonstrate a higher deposition efficiency, as illustrated in Figure 20. Surface roughness on the other hand stays around 2 µm as can be seen in Figure 19. This is remarkable and may be due to the fact, that the heavy particles do not change their velocity significantly due to their high inertia at higher spray distances.

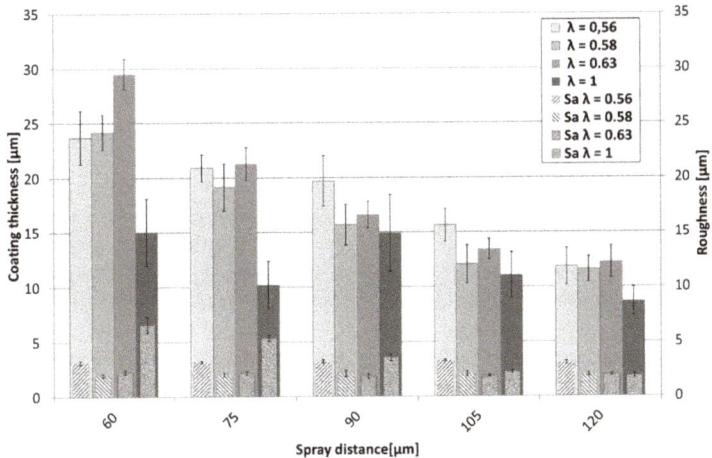

Figure 19. Comparison of coating thickness and roughness values of Ag coatings sprayed at different spray distances and λ values.

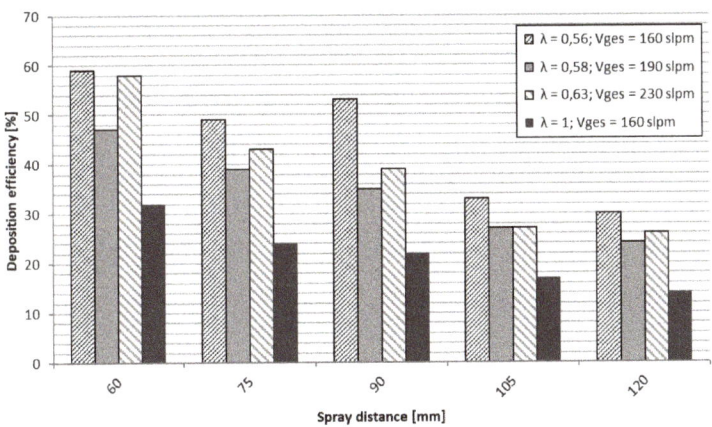

Figure 20. Deposition efficiency of Ag coatings for various spray parameter sets. Different spray distances for given sets of λ and total gas volume (slpm).

Spraying with stoichiometric parameter (λ = 1) on the other hand leads to the lowest deposition rate/efficiency and highest roughness values. The high roughness is mainly explained by the formation of defects on the coating surface as already discussed. This can be also clearly observed from the cross-section in Figure 16 (third row). Apparently, this parameter set is unsuitable to deposit defect free and smooth coatings. So far, the role of oxide formation at higher spray distances could not be clearly resolved in this study.

4. Conclusions and Outlook

In this study, suspensions containing metal powders of NiCr 80/20, copper and silver, suitable for the HVSFS process, were successfully prepared, characterized, and sprayed. Through variation of ethene and oxygen gas flow as well as the spraying distance, a suitable parameter window for each material could be established. The aim to spray dense and thin coatings with low oxidation rate and high metal content was fully achieved for silver and at least partially for copper. The results can open up interesting possibilities for electrical applications. Thin silver coatings may also serve as antibacterial surfaces. However, reduced spraying distances are necessary to achieve defect free silver coatings with reasonable deposition efficiencies. To minimize heat flux to the substrate, a different torch type or modified torch concept with reduced thermal power may be advantageous. In a future study, an in depth characterization regarding electrical conductivity together with the role of oxide formation, should be carried out in more detail.

The NiCr coatings show a fine dispersion of metal and oxide phases. Microstructure and oxidation are less affected by the spraying distance in comparison to silver and copper but can be adjusted by a variation of the gas flow. Due to its fine and evenly distributed oxide and the possibility to adjust coating thickness down to 10 µm or even lower, suspension sprayed NiCr can be an interesting approach for the improvement in film heating applications. This approach has been described and published recently by some of the authors [26].

Suspension preparation for metal powders on the other hand stays as a demanding task. High solid content and a long-term stability is overall difficult to achieve. As long as sedimentation occurs without the forming of stable aggregates, the suspension can stay processible by means of agitation and redispersion. This was especially the case for the suspensions used in this study. To reach industrially appropriate suspensions however, further development is necessary at this point.

Author Contributions: M.B. and P.K.: Experimental work at University of Stuttgart; A.K.: Research project leader (TherMes) at University of Stuttgart, J.L.: experimental work at obz innovation GmbH; F.T.: Research project leader (TherMes) at obz innovation GmbH. R.G.: Institute director IFKB University of Stuttgart. All authors have read and agreed to the published version of the manuscript.

Funding: This research project was supported by the German Ministry of Economic Affairs within the framework of a project (KF2121015AG4 TherMes) in the Central Innovation Program for Small and Medium-Sized Enterprises (ZIM).

Conflicts of Interest: The authors declare no conflict of interest.

References

1. Pawlowski, L. Suspension and Solution Thermal Spray Coatings. *Surf. Coatings Tech.* **2009**, *203*, 2807–2829. [CrossRef]
2. Toma, F.L.; Berger, L.M.; Scheiz, S.; Langner, S.; Rödel, C.; Potthoff, A.; Sauchuk, V.; Kusnezoff, M. Comparison of the Microstructural Characteristics and Electrical Properties of Thermally Sprayed Al2O3 Coatings from Aqueous Suspensions and Feedstock Powders. *J. Therm. Spray Tech.* **2012**, *21*, 480–488. [CrossRef]
3. Bolelli, G.; Cannillo, V.; Gadow, R.; Killinger, A.; Lusvarghi, L.; Rauch, J. Properties of High Velocity Suspension Flame Sprayed (HVSFS) TiO2 coatings. *Surf. Coatings Tech.* **2009**, *203*, 1722–1732. [CrossRef]
4. Toma, F.L.; Berger, L.M.; Jacquet, D.; Wicky, D.; Villaluenga, I.; de Miguel, Y.R.; Lindeløv, J.S. Comparative study on the photocatalytic behaviour of titanium oxide thermal sprayed coatings from powders and suspensions. *Surf. Coatings Tech.* **2009**, *203*, 2150–2156. [CrossRef]
5. Oberste Berghaus, J.; Legoux, J.G.; Moreau, C.; Hui, R.; Decès-Petit, C.; Qu, W.; Yick, S.; Wang, Z.; Maric, R.; Ghosh, D. Suspension HVOF Spraying of Reduced Temperature Solid Oxide Fuel Cell Electrolytes. *J. Therm. Spray Tech.* **2008**, *17*, 700–707. [CrossRef]
6. Bolelli, G.; Rauch, J.; Cannillo, V.; Killinger, A.; Lusvarghi, L.; Gadow, R. Investigation of High-Velocity Suspension Flame Sprayed (HVSFS) glass coatings. *Mater. Lett.* **2008**, *62*, 2772–2775. [CrossRef]

7. Stiegler, N.; Bellucci, D.; Bolelli, G.; Cannillo, V.; Gadow, R.; Killinger, A.; Lusvarghi, L.; Sola, A. High-Velocity Suspension Flame Sprayed (HVSFS) Hydroxyapatite Coatings for Biomedical Applications. *J. Therm. Spray Tech.* **2012**, *21*, 275–287. [CrossRef]
8. Díaz, L.A.; Cabal, B.; Prado, C.; Moya, J.S.; Torrecillas, R.; Fernández, A.; Arhire, I.; Krieg, P.; Killinger, A.; Gadow, R. High-velocity suspension flame sprayed (HVSFS) soda-lime glass coating on titanium substrate: Its bactericidal behaviour. *J. Eur. Ceram. Soc.* **2016**, *36*, 2653–2658. [CrossRef]
9. Aghasibeig, M.; Mousavi, M.; Ben Ettouill, F.; Moreau, C.; Wuthrich, R.; Dolatabadi, A. Electrocatalytically Active Nickel-Based Electrode Coatings Formed by Atmospheric and Suspension Plasma Spraying. *J. Therm. Spray Tech.* **2014**, *23*, 220–226. [CrossRef]
10. Oberste Berghaus, J.; Marple, B.; Moreau, C. Suspension Plasma Spraying of Nanostructured WC-12Co Coatings. *J. Therm. Spray Tech.* **2006**, *15*, 676–681. [CrossRef]
11. Ma, X.Q.; Roth, J.; Gandy, D.W.; Frederick, G.J. A New High-Velocity Oxygen Fuel Process for Making Finely Structured and Highly Bonded Inconel Alloy Layers from Liquid Feedstock. *J. Therm. Spray Tech.* **2006**, *15*, 670–675. [CrossRef]
12. Toma, F.L.; Berger, L.M.; Naumann, T. Suspensionsspritzen—Das Potential einer neuen Spritztechnologie, Suspension Spraying—The Potential of a New Spray Technology. *Therm. Spray Bull.* **2010**, *62*, 24–29.
13. Ak, N.F.; Tekmen, C.; Ozdemir, I.; Soykan, H.S.; Celik, E. NiCr coatings on stainless steel by HVOF technique. *Surf. Coatings Tech.* **2003**, *174–175*, 1070–1073. [CrossRef]
14. Higuera, V.; Belzunce, F.J.; Carriles, A.; Poveda, S. Influence of the thermal-spray procedure on the properties of a nickel-chromium coating. *J. Mater. Sci.* **2002**, *37*, 649–654. [CrossRef]
15. Donner, K.R.; Gaertner, F.; Klassen, T. Metallization of Thin Al_2O_3 Layers in Power Electronics Using Cold Gas Spraying. *J. Therm. Spray Tech.* **2011**, *20*, 299–306. [CrossRef]
16. Chavan, N.M.; Ramakrishna, M.; Phani, P.S.; Rao, D.S.; Sundararajan, G. The influence of process parameters and heat treatment on the properties of cold sprayed silver coatings. *Surf. Coatings Tech.* **2011**, *205*, 4798–4807. [CrossRef]
17. Rolland, G.; Sallamand, P.; Guipont, V.; Jeandin, M.; Boller, E.; Bourda, C. Damage Study of Cold-Sprayed Composite Materials for Application to Electrical Contacts. *J. Therm. Spray Tech.* **2012**, *21*, 758–772. [CrossRef]
18. Krieg, P.; Killinger, A.; Gadow, R.; Burtscher, S.; Bernstein, A. High velocity suspension flame spraying (HVSFS) of metal doped bioceramic coatings. *Bioact. Mater.* **2017**, *2*, 162–169. [CrossRef]
19. Chen, Y.; Zheng, X.; Xie, Y.; Ding, C.; Ruan, H.; Fan, C. Anti-bacterial and cytotoxic properties of plasma sprayed silver-containing HA coatings. *J. Mater. Sci. Mater. Med.* **2008**, *19*, 3603–3609. [CrossRef]
20. Trenkle, F.; Winkelmann, M.; Wüst, F.; Luth, J. Thermal spraying of thin metallic coatings. *Therm. Spray Bull.* **2017**, *69*, 22–27.
21. Krieg, P. Dissertation Peter Valentin Krieg, Hochgeschwindigkeitssuspensionsflamm¬spritzen von Metallen und metalldotierten Biokeramiken. In *Forschungsberichte des Instituts für Fertigungs—technologie keramischer Bauteile (IFKB)*; Gadow, R., Ed.; Shaker Verlag: Aachen, Germany, 2018; ISBN 978-3-8440-6091-1.
22. Fauchais, P.L.; Heberlein, J.V.R.; Boulos, M.I. *Thermal Spray Fundamentals: From Powder to Part*; Springer: Boston, MA, USA, 2014; pp. 1028–1029. [CrossRef]
23. Brousse, E.; Montavon, G.; Fauchais, P.; Denoirjean, A.; Rat, V.; Coudert, J.-F.; Ageorges, H. Thin and dense yttria-partially stabalized zirconia electrolytes for IT-SOFC manufactured by suspension plasma spraying. *Therm. Spray Crossing Bord. (DVS)*, 2008; 547–552.
24. Racek, O. The effect of HVOF particle-substrate interactions on local variations in the coating microstructure and the corrosion resistance. *J. Therm. Spray Tech.* **2010**, *19*, 841–851. [CrossRef]
25. Förg, A.; Blum, M.; Killinger, A.; Nicolás, J.A.M.; Gadow, R. Deposition of chromium oxide-chromium carbide coatings via high velocity suspension flame spraying (HVSFS). *Surf. Coatings Tech.* **2018**, *351*, 171–176. [CrossRef]
26. Luth, J.; Winkelmann, M.; Wüst, F.; Hartmann, S.; Trenkle, F.; Krieg, P.; Killinger, A. Dünne, metallische Heizschichten, hergestellt durch Suspensionsspritzen - Thin, metallic heating coatings, manufactured by means of suspension spraying. *Therm. Spray Bull.* **2019**, *12*, 26.

© 2020 by the authors. Licensee MDPI, Basel, Switzerland. This article is an open access article distributed under the terms and conditions of the Creative Commons Attribution (CC BY) license (http://creativecommons.org/licenses/by/4.0/).

Article

Performance of Hybrid Powder-Suspension Axial Plasma Sprayed Al$_2$O$_3$—YSZ Coatings in Bovine Serum Solution

Vasanth Gopal [1,2,*], Sneha Goel [3], Geetha Manivasagam [2] and Shrikant Joshi [3]

1 Department of Physics, School of Advanced Sciences, VIT, Vellore 632014, India
2 Centre for Biomaterials, Cellular and Molecular Theranostics, VIT, Vellore 632014, India; geethamanivasagam@vit.ac.in
3 Department of Engineering sciences, University West, 46186 Trollhättan, Sweden; sneha.goel@hv.se (S.G.); shrikant.joshi@hv.se (S.J.)
* Correspondence: vasant.phy87@gmail.com; Tel.: +91-869-554-1071

Received: 24 April 2019; Accepted: 10 June 2019; Published: 14 June 2019

Abstract: Ceramic coatings on metallic implants are a promising alternative to conventional implants due to their ability to offer superior wear resistance. The present work investigates the sliding wear behavior under bovine serum solution and indentation crack growth resistance of four coatings, namely (1) conventional powder-derived alumina coating (Ap), (2) suspension-derived alumina coating (As), (3) composite Al$_2$O$_3$—20wt % Yittria stabilized Zirconia (YSZ) coating (AsYs) deposited using a mixed suspension, and (4) powder Al$_2$O$_3$—suspension YSZ hybrid composite coating ApYs developed by axial feeding plasma spraying, respectively. The indentation crack growth resistance of the hybrid coating was superior due to the inclusion of distributed fine YSZ particles along with coarser alumina splats. Enhanced wear resistance was observed for the powder derived Ap and the hybrid ApYs coatings, whereas the suspension sprayed As and AsYs coatings significantly deteriorated due to extensive pitting.

Keywords: axial feeding; hybrid plasma spray coating; bovine serum solution; sliding wear; indentation

1. Introduction

Thermal spray technique is one of the most versatile surface engineering processes and enables the production of protective coatings, especially for harsh environments such as wear, corrosion/oxidation, high temperature, etc. Among the different variants within the thermal spray family, plasma spraying is a well-established technique used in many applications, such as aerospace, automotive, marine, and medical devices [1,2]. Particularly, plasma spraying has an important role in orthopaedic implant application, where it is used to develop bio-ceramic (hydroxyapatite) coatings on the hip femoral stem to enhance Osseo-integration. Furthermore, the hydroxyapatite coating on the hip femoral stem is approved by the Food and Drug Administration (FDA) in the USA [3,4]. Plasma spraying involves the use of high energy plasma to transform the powder particles of the coating material into their molten state, and these are then propelled towards the substrate to form a coating. Usually, conventional plasma spraying uses a powder feedstock with a particle size distribution in the range of 10–100 μm, which results in a relatively coarse structured coating. Recently, the deposition of finely structured coatings using plasma spraying has gained attention due to their ability to yield improved performance [5–7]. However, the production of such finely structured coatings using plasma spraying has traditionally been a huge challenge due to the inherent flowability problems associated with the use of fine feedstock powders [8].

Suspension plasma spraying (SPS) has now emerged as an attractive technique to produce fine structured coatings. The process involves suspending a fine sub-micron or nanometric powder in a solvent and injecting the resulting suspension into a plasma flame to overcome the above issue of flowability associated with fine powders [9,10]. Axial suspension feeding is a recent advancement in the SPS process and involves the suspension being injected axially into the plasma flame rather than radially. This axial feeding enhances the enthalpy exchange between the plasma flame and the sprayed material [11] and allows more efficient utilization of the plasma energy that is essential to achieve industrially relevant throughputs. The advent of the SPS process with axial feeding provides an opportunity to combine coarse powder feedstock and a fine structured suspension feedstock to realize hybrid coatings with superior properties. In the present study, four different coatings were developed using axial suspension plasma spraying. These included a conventional monolithic spray grade powder derived alumina coating (Ap) and a pure suspension-derived alumina coating (As). These were compared with two composite alumina—yttria-stabilized zirconia (YSZ) coatings, one derived from a mixed suspension (AsYs) and the other deposited using a hybrid feedstock (ApYs), in which a conventional micron-sized alumina powder and fine sized YSZ in the form of suspension were axially injected simultaneously into the plasma flame. The coating developed by the hybrid process possesses a multi-scale structure, i.e., the presence of both micron- and fine-sized features, due to typical splat sizes that result from the conventional coarse spray-grade powder and the much finer powder that constitutes the suspension [12]. The choice of alumina and YSZ as an illustrative material system to demonstrate the efficacy of hybrid coatings was based on the combination of high hardness and Young's modulus of the Al_2O_3 matrix with an additional toughening effect expected from the YSZ dispersion, thereby significantly increasing the flexural strength and the fracture toughness of Al_2O_3 [13–15]. Furthermore, the application of zirconia toughened alumina in hip implants has been already reported to significantly reduce the failure rate of implants due to the superior wear resistance [16,17].

Although the zirconia toughened alumina hip implants have significantly reduced the failure rate, the sudden fracture of bulk ceramic implants has raised inevitable concerns over their utility. Consequently, ceramic coatings on metallic implants are a promising alternative due to their ability to offer superior wear resistance, with the metallic substrate providing good fracture toughness. Considering the impact of the wear and the fracture toughness on the longevity and the performance of hip implants, the present study aimed to investigate the sliding wear behaviour of all the above-mentioned ceramic coatings in the presence of a bovine serum albumin solution. In addition to wear resistance, the indentation crack growth resistance of the developed coatings was also investigated, and the prominent results are presented herein.

2. Experimental Methods

2.1. Substrate Materials, Powders, and Suspensions

The suspensions used in the study were of Al_2O_3 (d10 = 0.51 µm, d50 = 2.20 µm, d90 = 4.93 µm) for As and 8 wt % YSZ (d10 = 230 nm, d50 = 440 nm, d90 = 950 nm) for ApYs coating. A mixed suspension of the above two to obtain an Al_2O_3-20wt %YSZ composition was used in the case of the AsYs coating. All the suspensions were procured from the same supplier (Treibacher, Austria) with identical solid load content of 25 wt % in ethanol. Spray-grade powders of Al_2O_3 (AMPERIT® 740, fused and crushed, 22/5 µm, H.C. Starck GmbH, München, Germany) and NiCoCrAlY (AMPERIT® 410, gas atomized, H.C. Starck GmbH, München, Germany) were also used as feedstock materials. The substrate material employed was Domex 355 steel for all the coatings. Although this material is not specifically intended for biomedical applications, it was used out of convenience since, in the case of overlay coatings such as those that are plasma sprayed, the investigated wear resistance properties of the deposited layer are not expected to be influenced by the substrate.

2.2. Coating Deposition

All the coatings produced in this study were sprayed using a Mettech Axial III high power plasma torch (Northwest Mettech Corp., Vancouver, BC, Canada) capable of the axial introduction of feedstock and equipped with a Nanofeed 350 suspension feeder as well as a separate powder feed system (Uniquecoat, model PF50WL, Richmond, VA, USA). The plasma spray process involved in the current study has three different processes:

(a) The Axial feed of conventional powder to develop Ap coating. In this process, the feedstock powder was carried inside a tube and injected just before the plasma gun exit. Argon was used as a carrier gas to transform the powder from the powder feeder to the plasma torch.

(b) The second process involved the axial feed of suspension with a coaxial feed of atomizing gas to develop As and AsYs coatings. Instead of the simple tube in the case of Ap coating, a larger tube housing a coaxially mounted liquid feedstock injector was used in suspension spraying. The inner tube carried the suspension while surrounding the inner tube; an atomizing gas was fed into the annular region to atomize the feedstock. The liquid feedstock was pumped into the plasma torch, and the feed rate could be varied by varying the atomizing gas.

(c) The third process involved the combination of the above two processes to fetch the hybrid powder-suspension ApYs coating. The process involved the axial feed of suspension with a coaxial feed of powder with a carrier gas. The amount of suspension, powder feed rate, and carrier/atomizing gas could be controlled individually. The process parameters employed for depositing the coatings As, AsYs, Ap, and ApYs are given in Table 1. A schematic representation of the plasma spray process involved in the current study is depicted in Figure 1. Prior to the coating, the substrates were grit blasted with alumina grit of 80 µm size to induce surface roughness, which enhanced the mechanical interlocking between coating and substrate. Furthermore, the grit blasted substrates were the first bond coated with NiCoCrAlY to enhance the adhesion of the coatings.

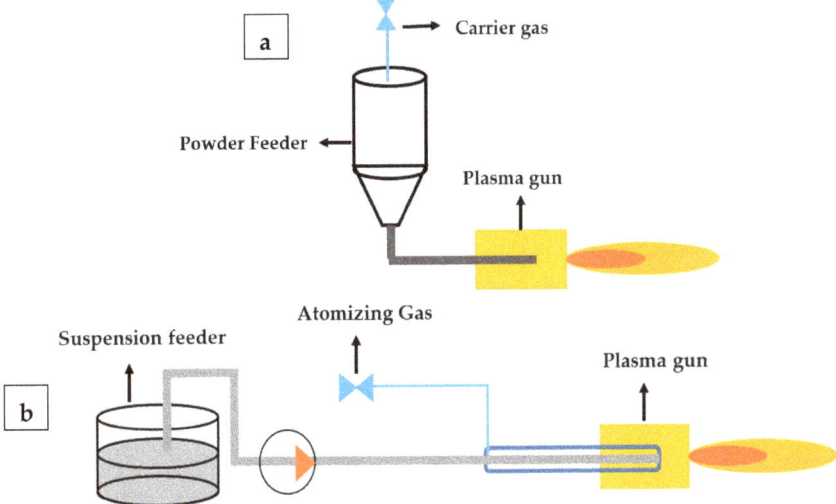

Figure 1. *Cont.*

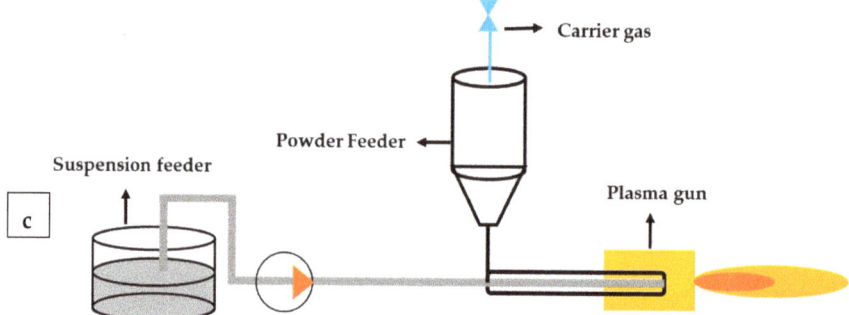

Figure 1. Schematic representation of the axial plasma spraying process of (a) Ap, (b) As and AsYs (c) ApYs coatings.

Table 1. Process parameters employed for depositing As, AsYs, Ap, and ApYs coatings.

Parameters	ApYs	Ap	As	AsYs
Operating gas	H_2, N_2	H_2, N_2	H_2, N_2, Ar	H_2, N_2, Ar
Spray Distance, mm	150	150	100	100
Current, A	230	230	220	220
Powder feed rate, g/min	40	50	-	-
Suspension feed rate, mL/min	40	-	40	40
Power, kW	124	109	122	122

2.3. Coating Characterization

Microstructural investigation of polished cross-sections of the coatings was carried out using SEM (HITACHI, TM3000, Tokyo, Japan). The phase constitution of the composite coatings was analyzed using a Siemens D500 XRD with Cr Kα radiation (λ = 2.29 Å) with the diffraction angle (2θ) varied between 10° and 120°. In addition, an energy dispersive spectroscopy (EDS) system (EDAX, Mahwah, NJ, USA) was utilized for composition analysis of cross-sections of the coating. The porosity of the coatings was measured using an image analysis technique ImageJ software (National Institute of Health, USA). Twenty cross-sectional SEM images were taken at appropriate magnification (×500) for porosity measurement. The grey scale images were converted into black and white by manual threshold adjustment. The average porosity values of the coatings were given in the table.

2.4. Mechanical Characterization

2.4.1. Vickers Micro Hardness

The hardness of all the coatings was measured using Vickers micro-hardness testing (HMV-2, Shimadzu Corp., Kyoto, Japan) with a constant load of 100 g and a dwell time of 15 s. A total of eight indents were recorded on each specimen, and the average hardness and the standard deviation were determined.

2.4.2. Indentation Crack Growth Resistance

The crack growth resistance of the coatings was evaluated by Vickers indentation method. The indentations were made on polished coating cross sections. The indenter was carefully positioned in such a way that the indent was made approximately at the centre of the coating. A load of 2 kg was chosen in the case of all the coatings, since lower loads were not found to lead to any discernible crack formation. A minimum of five indentations was carried out, and the lengths of the cracks were measured using an optical microscope (Carl Zeiss). The crack length C was measured using the equation [18]:

$$C = \frac{(2d_1 + 2d_2)}{4} + \frac{(l_1 + l_2)}{2}$$

where $2d_1$ and $2d_2$ are the parallel and the perpendicular Vickers indenter diagonals, l_1 and l_2 are the lengths of the cracks observed on either side of the indent parallel to the lamellae, depicted to originate below in Figure 2 from the left and the right side of the indent corners. A schematic representation of crack propagation from the Vickers indenter corner and the measurement of the crack length are shown in Figure 2 below:

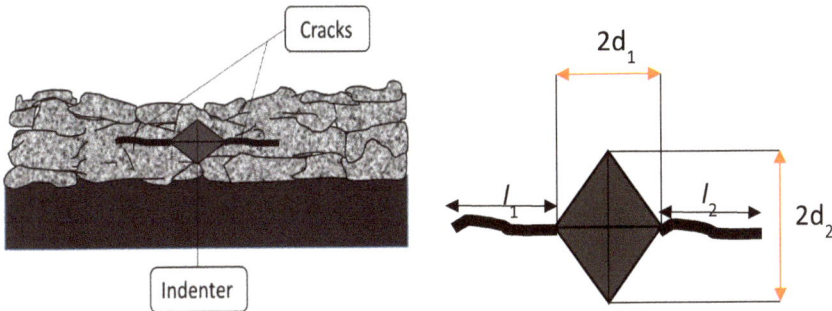

Figure 2. Schematic representation of crack propagation from Vickers indentation and the measurement of crack length.

2.5. Sliding Wear Test

The coatings were subjected to a sliding wear test carried out using a linearly reciprocating wear tester TR-285 M machine (DUCOM, Bengaluru, India) as per American Society for Testing and Materials (ASTM) standard G133-05 [19]. An alumina ball with a diameter of 6 mm was used as the counterpart. The wear tests were carried out in the presence of a 4 mg/mL bovine serum albumin solution with the addition of 0.2% sodium azide to retard bacterial growth. The whole test set up was maintained at 37 °C for 10^5 cycles. Prior to the wear test, all the coatings were polished using SiC emery sheets (600 to 3000 grit size) and finally polished with 0.1 µm diamond paste. The polished coatings were ultrasonically cleaned in acetone to get rid of any contaminants. The wear volume was calculated by following the ASTM G133-05(2016) standard [19] using the relationship $V = A \times l$, where A is the average cross-sectional area of the track (mm^2) and l length of the stroke (mm). The area of the wear track was measured using a three-dimensional (3-D) optical profilometer (Taylor Hobson, Leicester, UK). The specific wear rate was measured using the following formula:

$$WR = \frac{V}{L \times S}$$

where WR represents wear rate (mm^3 N^{-1} m^{-1}), V is the wear volume (mm^3) determined as the product of the mean worn area and the width of wear track, L corresponds to normal load (N), and S signifies sliding distance (m).

3. Results and Discussion

3.1. Cross-Sectional Morphology

Figure 3 shows the cross-sectional morphology of the four coatings, As, AsYs, Ap, and ApYs. All the coatings revealed a well-intact interface with the bond coat, which was indicative of good adhesion.

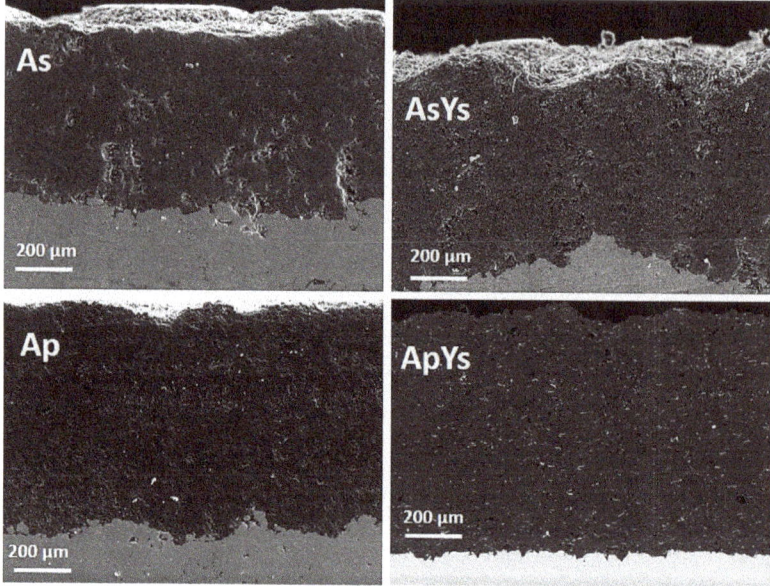

Figure 3. Cross-sectional SEM images of As, AsYs, Ap, and ApYs coatings.

Figure 4 shows higher magnification images of the coatings AsYs and ApYs, which revealed the incorporation of YSZ (seen as the brighter phase) in the coating over a large area. Particularly, the incorporation of YSZ was finer and more homogenous in the AsYs coating compared to the ApYs coating, plausibly due to the significantly smaller particle size of the Alumina splats comprising the matrix. In addition, EDS analysis was carried out on the cross-section of the ApYs coating to confirm the presence of the Zr region. Spot 1, which was taken at the darker area, confirmed the presence of Al_2O_3, whereas spot 2 at the brighter region confirmed the presence of a Zr rich region, as shown in Figure 5. The percentage porosities of all the coatings were determined by image analysis (using ImageJ software) and are given in Table 2. The porosity of the AsYs (3.5 ± 2.2 vol%) was found to be the highest among all the coatings, whereas the porosities of Ap, As, and ApYs coatings showed no significant difference.

Table 2. Percentage porosity values of the coatings.

Coatings	As	AsYs	Ap	ApYs
Porosity (Vol%)	2.5 ± 0.9	3.5 ± 1.2	2.2 ± 0.3	2.6 ± 0.3

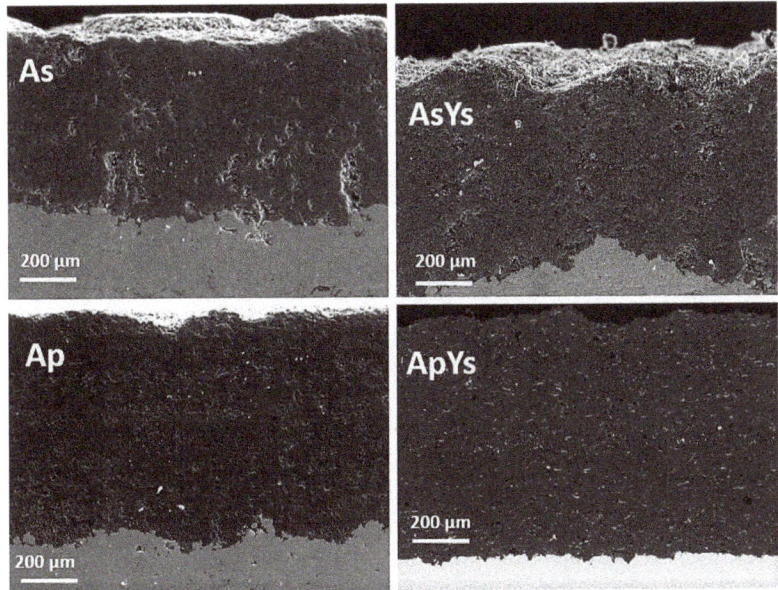

Figure 4. Cross-sectional SEM images of As, AsYs, Ap, and ApYs coatings taken at higher magnification.

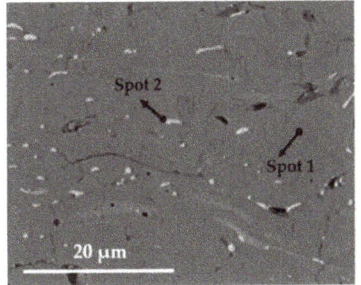

Spot	O	Al	Zr	Y
	wt.%	wt.%	wt.%	wt.%
1	54.0	46.0	-	-
2	29.9	0.9	65	4.2

Figure 5. Energy dispersive spectroscopy (EDS) analysis on the cross-sectional SEM image of hybrid ApYs coating and its elemental composition taken at two different spots confirms the bright region to be Zr.

3.2. XRD Analysis

The XRD patterns of the coatings (As, AsYs, Ap, and ApYs) are shown in Figure 6. The conventional alumina coating Ap exhibited the typical XRD pattern representative of conventional plasma sprayed alumina coatings in which the metastable γ-Al_2O_3 was observed to be the major phase along with traces of α-Al_2O_3 phase. It is well known that in conventional plasma spraying, the presence of γ-Al_2O_3 is attributable to the rapid splat quenching of alumina during the plasma spray process, whereas the α-Al_2O_3 results from partially melted or unmelted alumina particles [20–24]. On the contrary, the coating As which was produced by the SPS technique showed α-Al_2O_3 as the predominant phase rather than γ-Al_2O_3. The disparity in the alumina phase constitution between the conventional atmospheric plasma spraying and suspension spraying is that, in the conventional spraying, the alumina particles are in direct contact with the plasma flame, which subjects the particles to extensive melting and subsequent rapid splat quenching, leading to the formation of a significant amount of γ-Al_2O_3 phase. In the latter case, the evaporation of the solvent consumes a substantial portion of the energy available in the plasma flame, which plausibly results in lower particle temperature at impact, leading to reduced cooling rates that promote the retention of the α-Al_2O_3 phase. In this context, it is also relevant to mention that a prior study already demonstrated the formation of phase-pure α-Al_2O_3 using a liquid feedstock involving a solution precursor rather than a suspension [25]. In the case of both the all-suspension composite AsYs coating and the powder-suspension hybrid composite ApYs coating, the presence of tetragonal ZrO_2 distributed in Al_2O_3 was confirmed. Furthermore, all the coatings that involved the use of a suspension feedstock showed a broad hump in 2theta position from approximately 20° to 60°, which was suggestive of the presence of amorphous phases in the coatings.

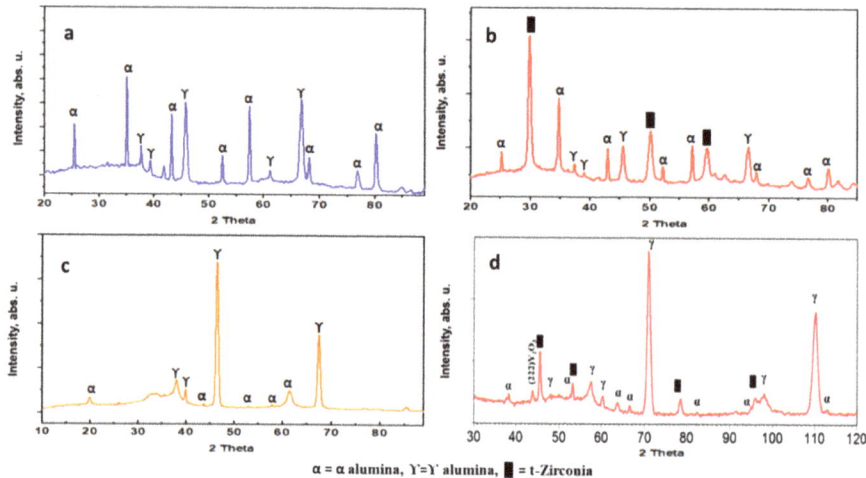

Figure 6. XRD spectra of coatings (**a**) As, (**b**) AsYs, (**c**) Ap, and (**d**) ApYs.

3.3. Hardness

Figure 7 shows the Vickers microhardness of As, AsYs, Ap, and ApYs coatings, respectively. Among all the powder-derived coatings, Ap exhibited the highest hardness of 1310 ± 15 HV followed by the hybrid coating ApYs 1021 ± 19 HV. Another interesting result was that the suspension coating As exhibited lower hardness compared to the conventional powder Ap coating. This lower hardness could be attributable to the inherent differences in the microstructures of the powder-derived and the suspension-derived coatings in terms of individual splat sizes, a number of splat boundaries, porosity and pore distribution, etc. It was also noted that the suspension derived coating AsYs exhibited the lowest hardness 938 ± 08 HV among all the coatings. A relative decrease in hardness in the cases of ApYs and AsYs compared to Ap and As respectively was due to the inclusion of YSZ, which has an intrinsic lower hardness than alumina [22].

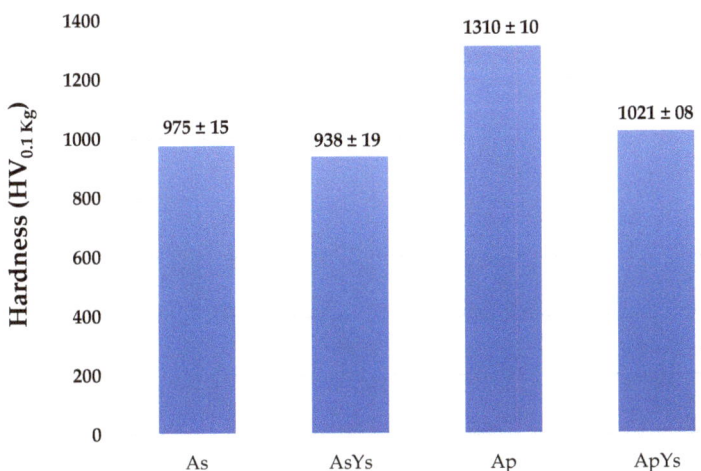

Figure 7. Vickers microhardness values of the coatings As, AsYs, Ap, and ApYs.

3.4. Indentation Crack Growth Resistance

The Vickers indentation technique is a versatile tool for assessing the fracture toughness of bulk brittle materials—particularly ceramics. However, the determination of fracture toughness of plasma sprayed coatings employing this technique is quite complicated due to the highly anisotropic microstructure associated with the typical lamellar structure [26]. Therefore, in the present work, the indentation technique was utilized to study the crack growth resistance of the coatings rather than for fracture toughness measurement. A similar approach of investigating the indentation crack growth resistance by measuring the crack length was reported previously by Hong Luo et al. [26]. Figure 8 shows the indentation crack growth in As, AsYs, Ap, and ApYs coatings. The crack length was measured and is tabulated in Table 3. From Figure 8, it is evident that the coatings As and Ap showed a long crack originating from the two indent corners parallel to the coating–substrate interface. However, there were no cracks observed from the indentation perpendicular to the coating–substrate interface. This asymmetric crack behaviour was due to the inherent splat microstructure of the plasma sprayed coatings [26]. On the contrary, the hybrid coating ApYs did not show any cracks around the indentation either parallel or perpendicular to the coating–substrate interface, whereas the coating AsYs exhibited short cracks along the plane of the coating–substrate interface.

Table 3. Indentation crack length in the coatings.

Coatings	Crack Length (μm)
As	111.35
AsYs	54.72
Ap	181.50
ApYs	No visible cracks

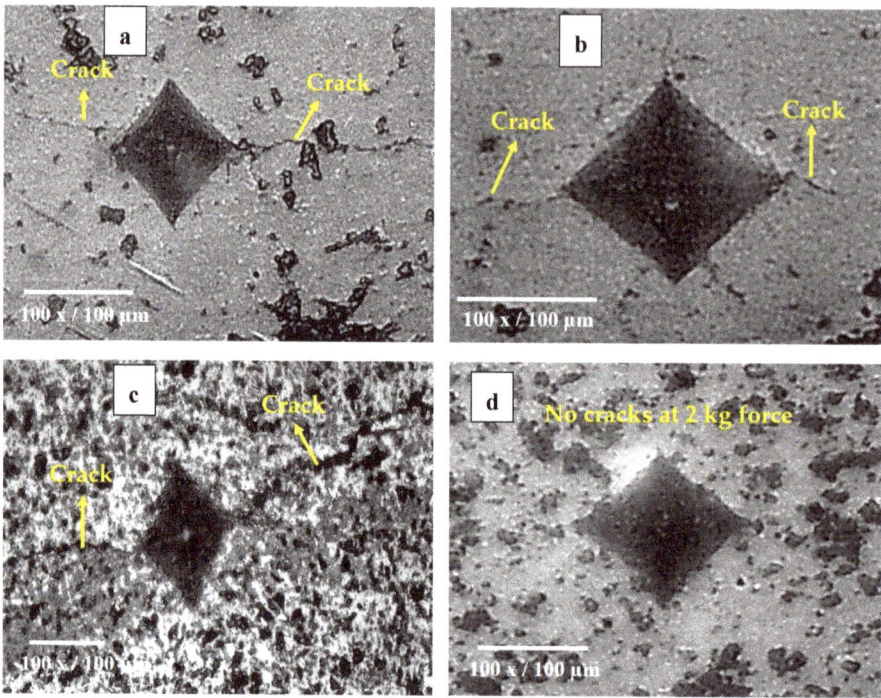

Figure 8. Optical micrographs revealing indentation crack growth in (**a**) As (**b**) AsYs (**c**) Ap and (**d**) ApYs.

The crack growth resistance of a coating was evaluated from the length of the cracks that emerged from the two indenter corners parallel to the coating–substrate interface. The smaller the crack length, the larger the crack growth resistance. In this context, the hybrid coating ApYs was observed to possess superior crack growth resistance with no crack originating from the indenter corners, even at 2 kg force. The enhanced crack growth resistance of the ApYs coating could plausibly be attributed to one or more of the following: (i) the intrinsic phase transformation toughening associated with tetragonal-to-monoclinic transition of zirconia phase, which restricted crack propagation [16], or (ii) the inclusion of finer YSZ particles in coarser alumina splats leading to an enhancement in cohesion strength between the splats and thereby resulting in superior crack growth resistance. In the case of AsYs coatings, considerably smaller cracks were observed to have emanated at the indenter corners compared to As, which highlights the role of the finer YSZ included in the former coating in enhancing the crack growth resistance. According to the measured crack length, the indentation crack growth resistance of the coatings could be ranked as ApYs > AsYs > As > Ap. The inferior crack growth resistance of the powder derived Ap was due to the presence of γ-Al_2O_3 phase. It has been reported that the γ-Al_2O_3 phase present in the high velocity oxy fuel (HVOF) sprayed alumina coating worsened the fracture toughness compared to the mixed α and γ-Al_2O_3 phases [27].

3.5. Sliding Wear

The sliding wear behaviours of As, AsYs, Ap, and ApYs coatings were investigated in the presence of bovine serum albumin solution. In order to understand the prevailing wear mechanisms, the coatings were grouped into different categories. The coatings Ap–As were grouped to study the wear performance of conventional powder-derived and fine structured suspension-derived coatings. Coatings Ap–ApYs were compared to evaluate the effects of the introduction of a fine structured second phase on the performance of a conventional coarse-structured Ap coating. Similarly, coatings As–AsYs were also compared to investigate how the introduction of fine structured YSZ second phase affected the performance of the fine structured As coating.

3.5.1. Specific Wear Rate

The specific wear rates of the coatings were calculated and are shown in Figure 9. The wear rate of the Ap coating was found to be the lowest among all the coatings. In contrast, the wear rate of the As coating was highest among all the coatings. In the case of the hybrid ApYs coating, the wear rate was almost similar to that of the Ap coatings. The wear rate of AsYs was slightly lower than As but significantly higher compared to Ap and ApYs coatings. Further, the wear depth assessed by 3-D optical profilometer in Figure 10, showed larger wear depth for coatings As and AsYs, whereas the wear depth was smaller for Ap and ApYs coatings. According to the wear rate values determined, the wear rate of the investigated coatings could be ranked in the order Ap ≈ ApYs < AsYs < As.

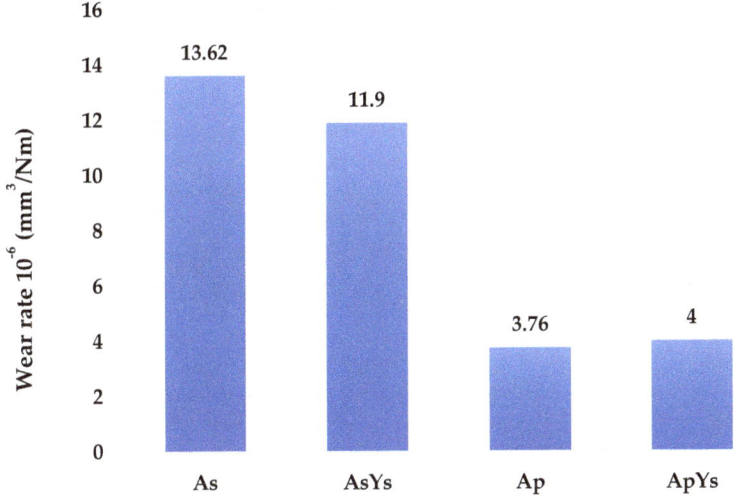

Figure 9. Specific wear rates of As, AsYs, Ap, and ApYs coatings.

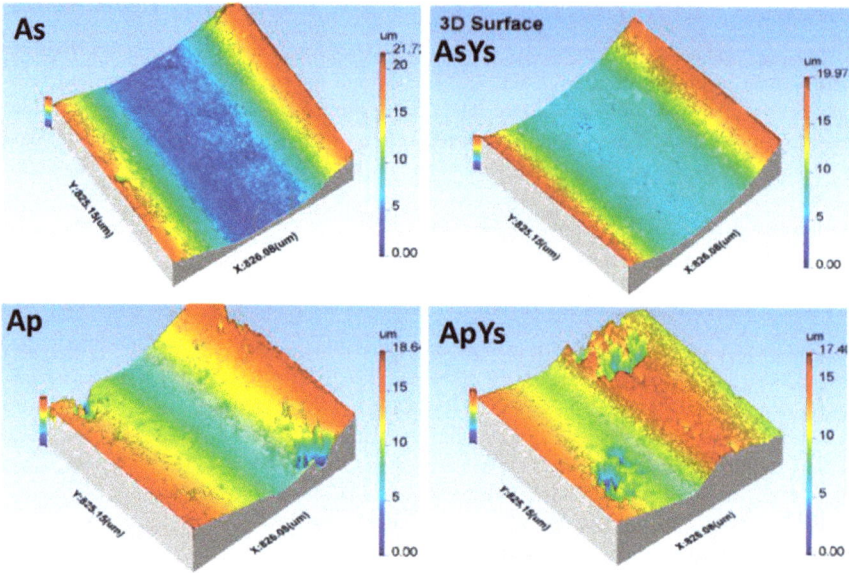

Figure 10. Three-dimensional (3-D)-optical profilometer images of coatings As, AsYs, Ap, and ApYs.

3.5.2. Worn Surface Morphology

Figure 11 shows the worn surface morphology of all the coatings. The worn surface of the Ap coating revealed a smooth surface. Higher magnification images showed uniformly distributed pits with no evidence of fine grooves or wear debris. Furthermore, in the wear track, underlying splats unaffected during sliding were also observed. In comparison, the worn surface morphology of the As coating revealed substantial pitting formation with a rougher surface. Furthermore, at higher magnification, the presence of nano-sized wear debris was also observed. When comparing ApYs and AsYs coatings, the latter showed extensive pitting with a rough, worn surface, whereas the ApYs coatings showed occasional pitting. From the above results, it could be inferred that the suspension derived coatings As and AsYs underwent severe pitting in serum solution, whereas the powder derived Ap and hybrid ApYs coatings exhibited occasional pitting. The extent of pitting directly influenced the wear resistance of the coatings. The coatings As and AsYs, which showed extensive pitting, possessed a higher wear rate; particularly, the As coating possessed the highest wear rate among all the coatings. In contrast, the wear rates of coatings Ap and ApYs were significantly reduced.

As suggested by Hawthrone et al. [28], the wear of thermal spray coatings occurs via three processes, (i) microchipping and ploughing, (ii) debonding at splat boundaries, and (iii) splat fractures associated with porosity. In the case of coatings with poor inter-splat bonding, the primary material removal mechanism is debonding at splat boundaries and results in an increased wear rate.

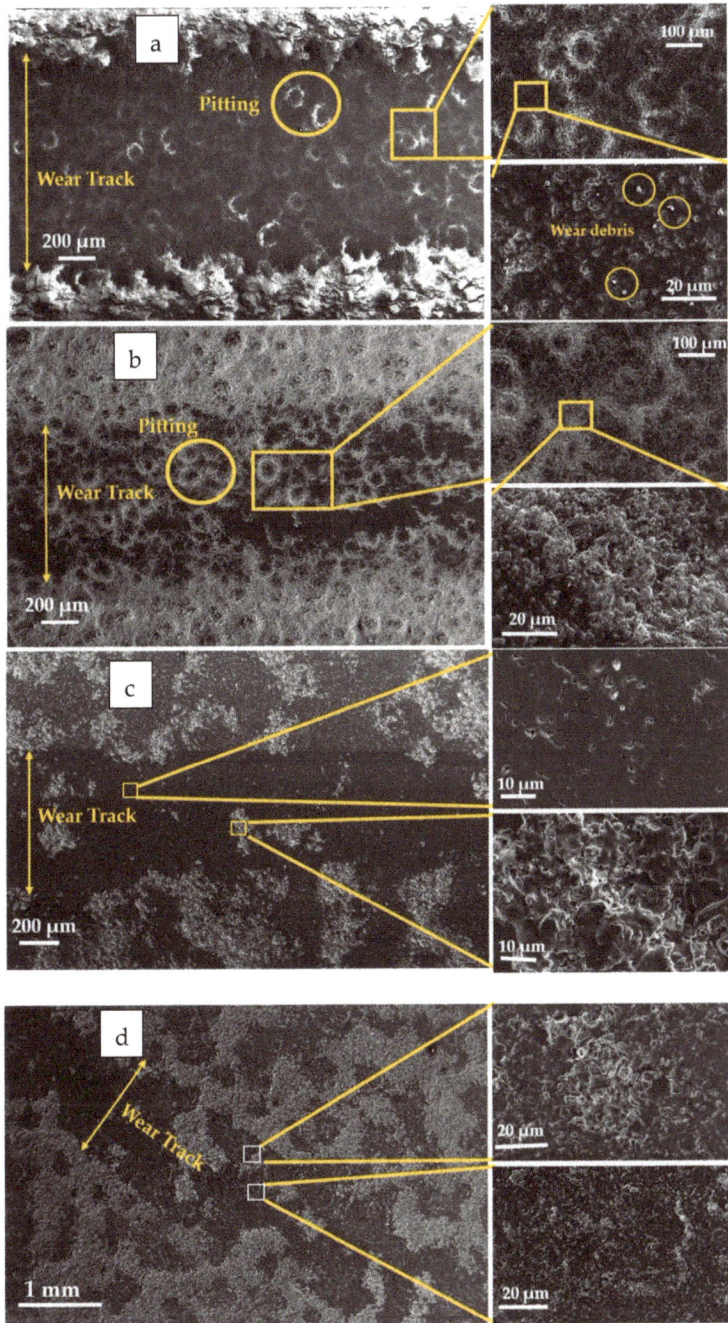

Figure 11. Worn surface morphologies at different magnifications in the case of (**a**) As, (**b**) AsYs, (**c**) Ap, and (**d**) ApYs coatings.

Additionally, many prior reports have shown that the wear resistance of plasma sprayed ceramic coatings increases with enhanced hardness and toughness [29–31]. In this context, the Ap coating was found to possess the highest hardness among all the coatings, which explains its superior wear resistance. It is noteworthy to mention that a previous study involving one of the authors showed that the coating Ap under dry sliding conditions possessed the highest specific wear rate among all the four coatings [32]. However, in the present study, the wear rate of the Ap coating was the least among all the coatings. This reduction in wear rate could be easily explained by the introduction of lubricant. Also, the presence of proteins in bovine serum solution could significantly influence the frictional and the wear behaviours. Many authors have reported that, under bovine serum lubrication, a tribolayer could be formed on the surface, which limits direct contact between mating materials, thereby reducing friction and wear [33,34]. Furthermore, the underlying splats that are not affected during sliding may act as reservoirs for the lubricant. During sliding, the lubricant tends to be drawn up to spread on the surface, leading to a reduction in friction, wear, and seizure [35]. This could also explain the absence of grooves and grain pull outs in the case of Ap. A schematic wear mechanism of all the coatings in the presence of bovine serum solution is shown in Figure 12.

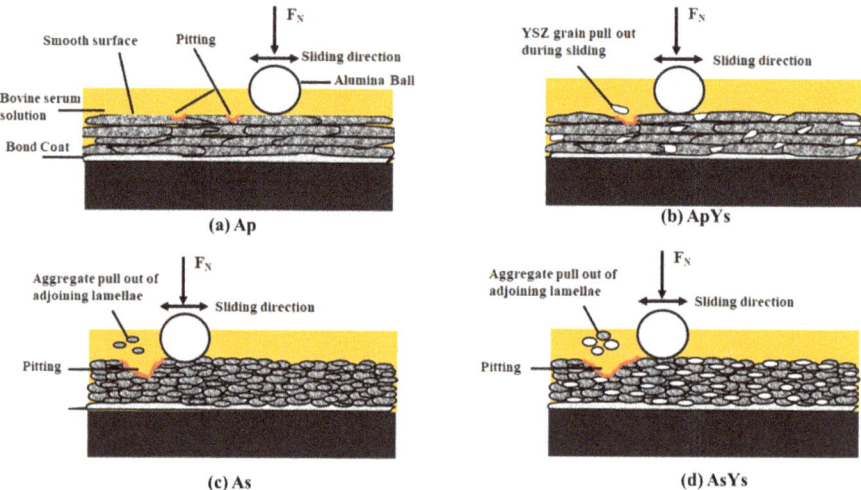

Figure 12. Schematic representation of wear of all the coatings in bovine serum solution.

In the case of the hybrid ApYs coating, even though the hardness was lower compared to that of the Ap coating, the indentation crack growth resistance was enhanced by the inclusion of much finer YSZ, with the splat sizes of alumina and YSZ in ApYs differing by nearly two orders of magnitude. This combination of optimum hardness and enhanced fracture toughness could significantly reduce the brittle fracture, thereby resulting in a wear rate that was comparable to that of the Ap coating. In contrast, the suspension derived coatings (As and AsYs) exhibited relatively poor wear resistance. Furthermore, the worn surface morphology of the As and the AsYs coatings suggested that pitting in the presence of the bovine serum solution was a predominant damage mechanism and could have resulted in severe wear. It is essential to mention that the powder derived Ap and hybrid ApYs coatings were also found to have undergone pitting, but the extent was much lower compared to the As and the AsYs coatings. The formation of pitting of ceramic prosthesis under bovine serum solution was previously reported by Rainforth et al. [36]. They reported that the Biolox Delta (Zirconia toughened alumina ceramic hip prosthesis) underwent severe pitting in serum solution when compared to ultra-pure water and Carboxy Methyl Cellulose (CMC)-Na solution as a lubricant. The initiation of pitting was primarily due to the intergranular fracture and the grain pull-out with the formation of a localized region of craters. Furthermore, they also observed that the pits were first initiated from the removal of zirconia grains followed by some of the alumina grains. Similarly, the coatings As and AsYs, which had lower hardness and toughness compared to Ap and ApYs coatings, might have been subjected to intergranular fracture and grain pull-out during sliding, which substantiated the formation of pitting. Moreover, the extensive pitting of AsYs compared to the As coating could have been attributed to the removal of zirconia grains followed by alumina grains. Furthermore, it was supposed that the difference in the splat size of As and AsYs, which was approximately ten orders lesser than the Ap and the ApYs coatings, attributed to the extensive pitting. During sliding, aggregate pull out of adjoining lamellae took place and led to extensive pitting of the As and the AsYs coatings. On the other hand, the Ap and the ApYs coatings possess superior wear resistance behaviours under bovine serum solution despite showing local pitting formation.

3.6. The Coefficient of Friction

The evolution of the coefficients of friction (CoF) of Ap, As, AsYs, and ApYs coatings during wear tests conducted in the presence of bovine serum are shown in Figure 13. As expected, the Ap and the hybrid ApYs coatings exhibited low CoF values compared to the As and the AsYs coatings. The average CoF values of the Ap and the ApYs coatings were 0.21 and 0.24, respectively. In the cases of As and AsYs, the CoFs were found to be higher at 0.34 and 0.36, respectively.

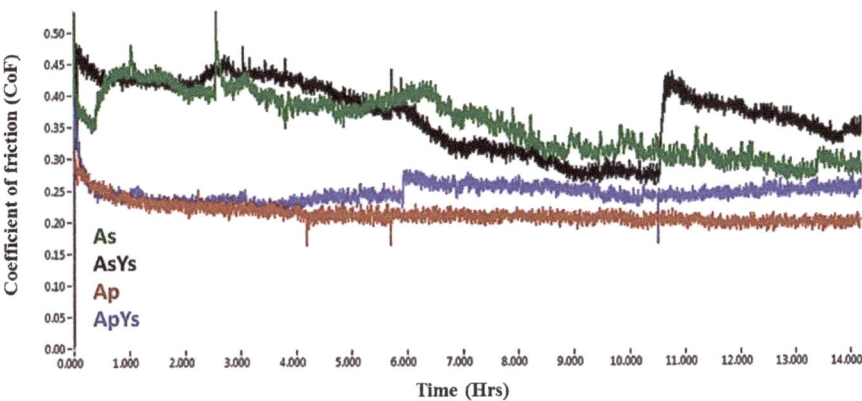

Figure 13. Coefficients of friction of As, AsYs, Ap, and ApYs coatings determined during wear testing in the presence of bovine serum.

The reason for lower CoF of the Ap coating was attributed to the microstructure of the coating. The conventional powder alumina coating yielded typically smooth, disk-like splats (fully molten splats), which usually resulted in lower local surface roughness. In addition, the polished, worn surface (see Figure 11c) and the presence of proteins in bovine serum solution also attributed to the lower CoF values of the Ap coating. Similarly, the hybrid coating ApYs also showed only marginally higher CoF values compared to the Ap coating despite showing a slightly rougher worn surface. In the case of the pure suspension coating As and the mixed suspension coating AsYs, the CoFs were found to be higher. As already seen in Section 3.5, the worn surface morphology of the pure suspension coatings (As and AsYs) revealed a rougher surface due to extensive pitting, which could explain the higher CoF values.

4. Conclusions

Alumina-YSZ ceramic composite coatings deposited using a hybrid powder-suspension feedstock (ApYs) were studied, and their microstructures and mechanical properties were compared with conventional powder-derived (Ap) and suspension-derived (As) alumina coatings, as well as coatings deposited using mixed alumina–YSZ suspension (AsYs). The sliding wear and the indentation crack growth resistance of these coatings were explicitly investigated, and the critical conclusions that could be drawn are as follows:

- The hardness of the conventional monolithic alumina coating Ap was the highest among all the coatings, which was 1.28 times higher than the hybrid ApYs coating, 1.34 times higher than the As coating, and 1.39 times higher than the AsYs coating. The inclusion of YSZ in both Ap and As coatings lowered the hardness by virtue of its lower intrinsic hardness.
- The indentation crack growth resistance of the hybrid coating ApYs was superior when compared to all the other coatings. The inclusion of finer YSZ particles into the coarser alumina splat increased the cohesion strength and resulted in superior crack growth resistance.
- The worn surface morphology of the Ap coatings exhibited polishing wear without any grooves, whereas the hybrid coating ApYs coating revealed local pitting. In the case of the pure and the mixed suspension coatings (As and AsYs), extensive pitting was observed, which deteriorated the wear resistance properties.

Overall, the hybrid coating ApYs showed promising results in terms of superior indentation crack growth resistance without any compromise in wear resistance properties compared to the other coatings. The optimum wear resistance and the superior indentation crack growth resistance of the hybrid ApYs coating could be a potential candidate for implant application. On the other hand, the suspension derived coatings (As and AsYs) showed detrimental properties by exhibiting severe pitting during sliding under the bovine serum solution.

Author Contributions: Conceptualization, S.J.; Methodology, V.G., S.G. and S.J.; Validation, V.G., G.M. and S.J.; Formal Analysis, G.M.; Investigation, V.G. and G.M.; Resources, V.G. and S.J.; Writing-Original Draft Preparation, V.G.; Writing-Review & Editing, V.G. and S.J.

Funding: This research received no external funding.

Acknowledgments: The authors wish to thank Stefan Björklund, research engineer, University West, Sweden for his assistance in plasma spraying. The authors Geetha Manivasagam and Vasanth Gopal sincerely thank VIT, Vellore for providing facilities to carry out the experiments.

Conflicts of Interest: The authors declare no conflict of interest.

References

1. Saber-Samandari, S.; Berndt, C.C. IFTHSE Global 21: Heat treatment and surface engineering in the twenty-first century Part 10–Thermal spray coatings: A technology review. *Int. Heat Treat. Surf. Eng.* **2013**, *4*, 7–13. [CrossRef]
2. Plasma Spray Coating Thermal Spray Coating Thermal Spray Technologies Inc. Available online: http://www.tstcoatings.com/plasma_spray.html (accessed on 14 June 2016).
3. Ivanka, I.; Vladislav, A.; Christoph, M.S.; Hristo, K.S.; Boyko, G. Plasma Sprayed Bioceramic Coatings on Ti-Based Substrates: Methods for Investigation of Their Crystallographic Structures and Mechanical Properties. In *Advanced Plasma Spray Applications*; IntechOpen: London, UK, 2012; Available online: https://www.intechopen.com/books/advanced-plasma-spray-applications/plasma-sprayed-bioceramic-coatings-on-ti-based-substrates-methods-for-investigation-of-their-crystal (accessed on 21 March 2012). [CrossRef]
4. Chu, P.K.; Chen, J.Y.; Wang, L.P.; Huang, N. Plasma-surface modification of biomaterials. *Mater. Sci. Eng. R.* **2002**, *36*, 143–206. [CrossRef]
5. Zhou, C.; Wang, N.; Wang, Z.; Gong, S.; Xu, H. Thermal cycling life and thermal diffusivity of a plasma-sprayed nanostructured thermal barrier coating. *Scr. Mater.* **2004**, *51*, 945–948. [CrossRef]
6. Chen, H.; Zhou, X.; Ding, C. Investigation of the thermomechanical properties of a plasma-sprayed nanostructured zirconia coating. *J. Eur. Ceram. Soc.* **2003**, *23*, 1449–1455. [CrossRef]
7. Liang, B.; Ding, C. Thermal shock resistances of nanostructured and conventional zirconia coatings deposited by atmospheric plasma spraying. *Surf. Coat. Technol.* **2005**, *197*, 185–192. [CrossRef]

8. Brinley, E.; Babu, K.S.; Seal, S. The solution precursor plasma spray processing of nanomaterials. *JOM* **2007**, *59*, 54–59. [CrossRef]
9. Ganvir, A. Comparative Analysis of Thermal Barrier Coatings Produced Using Suspension and Solution Precursor Feedstock. Master's Thesis, University West, Trollhättan, Sweden, 2014.
10. Fauchais, P.; Montavon, G.; Lima, R.S.; Marple, B.R. Engineering a new class of thermal spray nano-based microstructures from agglomerated nanostructured particles, suspensions and solutions: An invited review. *J. Phys. Appl. Phys.* **2011**, *44*, 93001. [CrossRef]
11. He, P.; Sun, H.; Gui, Y.; Lapostolle, F.; Liao, H.; Coddet, C. Microstructure and properties of nanostructured YSZ coating prepared by suspension plasma spraying at low pressure. *Surf. Coat. Technol.* **2015**, *261*, 318–326. [CrossRef]
12. Björklund, S.; Goel, S.; Joshi, S. Function-dependent coating architectures by hybrid powder-suspension plasma spraying: Injector design, processing and concept validation. *Mater. Des.* **2018**, *142*, 56–65. [CrossRef]
13. Chevalier, J.; Aza, A.H.D.; Fantozzi, G.; Schehl, M.; Torrecillas, R. Extending the Lifetime of Ceramic Orthopaedic Implants. *Adv. Mater.* **2000**, *12*, 1619–1621. [CrossRef]
14. Kern, F.; Palmero, P. Microstructure and mechanical properties of alumina 5 vol% zirconia nanocomposites prepared by powder coating and powder mixing routes. *Ceram. Int.* **2013**, *39*, 673–682. [CrossRef]
15. Mangalaraja, R.V.; Chandrasekhar, B.K.; Manohar, P. Effect of ceria on the physical, mechanical and thermal properties of yttria stabilized zirconia toughened alumina. *Mater. Sci. Eng. A* **2003**, *343*, 71–75. [CrossRef]
16. Kurtz, S.; Kocago, S.; Arnholt, C.; Huet, R.; Ueno, M.; Walter, W.L. Advances in zirconia toughened alumina biomaterials for total joint replacement. *J. Mech. Behav. Biomed. Mater.* **2014**, *31*, 107–116. [CrossRef] [PubMed]
17. Perrichon, A.; Haochih Liu, B.; Chevalier, J.; Gremillard, L.; Reynard, B.; Farizon, F.; Der Liao, J.; Geringer, J. Ageing, shocks and wear mechanism in ZTA and the long-term performance of hip joint materials. *Materials* **2017**, *10*, 569. [CrossRef] [PubMed]
18. Lima, M.M.; Godoy, C.; Modenesi, P.J.; Avelar-Batista, J.C.; Davison, A.; Matthews, A. Coating fracture toughness determined by Vickers indentation: An important parameter in cavitation erosion resistance of WC–Co thermally sprayed coatings. *Surf. Coat. Technol.* **2004**, *177–178*, 489–496. [CrossRef]
19. *Standard Test Method for Linearly Reciprocating Ball-on-Flat Sliding Wear*; ASTM G133-05(2016); ASTM International: West Conshohocken, PA, USA, 2016; Available online: www.astm.org (accessed on 15 April 2015).
20. Zhao, X.; An, Y.; Chen, J.; Zhou, H.; Yin, B. Properties of Al_2O_3–40 wt%ZrO_2 composite coatings from ultra-fine feedstocks by atmospheric plasma spraying. *Wear* **2008**, *265*, 1642–1648. [CrossRef]
21. Suffner, J.; Sieger, H.; Hahn, H.; Dosta, S.; Cano, I.G.; Guilemany, J.M.; Klimczyk, P.; Jaworska, L. Microstructure and mechanical properties of near-eutectic ZrO_2–60 wt%Al_2O–produced by quenched plasma spraying. *Mater. Sci. Eng. A* **2009**, *506*, 180–186. [CrossRef]
22. Dejang, N.; Limpichaipanit, A.; Watcharapasorn, A.; Wirojanupatump, S.; Niranatlumpong, P.; Jiansirisomboon, S. Fabrication and properties of plasma-sprayed Al_2O_3/ZrO_2 composite coatings. *J. Therm. Spray Technol.* **2011**, *20*, 1259–1268. [CrossRef]
23. McPherson, R. On the formation of thermally sprayed alumina coatings. *J. Mater. Sci.* **1980**, *15*, 3141–3149. [CrossRef]
24. Chen, D.; Jordan, E.H.; Gell, M. Microstructure of suspension plasma spray and air plasma spray Al_2O_3–ZrO_2 composite coatings. *J. Therm. Spray Technol.* **2009**, *18*, 421–426. [CrossRef]
25. Sivakumar, G.; Dusane, R.O.; Joshi, S.V. A novel approach to process phase pure α-Al_2O_3 coatings by solution precursor plasma spraying. *J. Eur. Ceram. Soc.* **2013**, *33*, 2823–2829. [CrossRef]
26. Luo, H.; Goberman, D.; Shaw, L.; Gell, M. Indentation fracture behavior of plasma-sprayed nanostructured Al_2O_3-13wt.%TiO_2 coatings. *Mater. Sci. Eng. A* **2003**, *346*, 237–245. [CrossRef]
27. Murray, J.W.; Ang, A.S.M.; Pala, Z.; Shaw, E.C.; Hussain, T. Suspension High Velocity Oxy-Fuel (SHVOF)-Sprayed Alumina Coatings: Microstructure, Nanoindentation and Wear. *J. Therm. Spray Technol.* **2016**, *25*, 1700–1710. [CrossRef]
28. Hawthorne, H.M.; Erickson, L.C.; Ross, D.; Tai, H.; Troczynski, T. The microstructure dependence of wear and indentation behaviour of some plasma-sprayed alumina coatings. *Wear* **1997**, *203–204*, 709–714. [CrossRef]
29. Erickson, L.C.; Hawthorne, H.M.; Troczynski, T. Correlations between micro- structural parameters, micro mechanical properties and wear resistance of plasma sprayed ceramic coatings. *Wear* **2001**, *250*, 569–575. [CrossRef]

30. Normand, B.; Fervel, V.; Coddet, C.; Nikitine, V. Tribological properties of plasma sprayed alumina–titania coatings: Role and control of the microstructure. *Surf. Coat. Technol.* **2000**, *123*, 278–287. [CrossRef]
31. Fervel, V.; Normand, B.; Coddet, C. Tribological behaviour of plasma sprayed Al_2O_3-based cermet coatings. *Wear* **1999**, *230*, 70–77. [CrossRef]
32. Murray, J.W.; Leva, A.; Joshi, S.; Hussain, T. Microstructure and wear behaviour of powder and suspension hybrid Al_2O_3-YSZ coatings. *Ceram. Int.* **2018**, *44*, 8498–8504. [CrossRef]
33. Yan, Y.; Yang, H.; Wang, L.; Su, Y.; Qiao, L. Effect of tribology processes on adsorption of albumin. *Surf. Topogr. Metrol. Prop.* **2016**, *4*, 014007. [CrossRef]
34. Parkes, M.; Myant, C.; Cann, P.M.; Wong, J.S.S. Synovial fluid lubrication: The effect of protein interactions on adsorbed and lubricating films. *Biotribology* **2015**, *1*, 51–60. [CrossRef]
35. Ahmed, A.; Masjuki, H.H.; Varman, M.; Kalam, M.A.; Habibullah, M.; Al Mahmud, K.A.H. An overview of geometrical parameters of surface texturing for piston/cylinder assembly and mechanical seals. *Meccanica* **2016**, *51*, 9–23. [CrossRef]
36. Ma, L.; Rainforth, W.M. The effect of lubrication on the friction and wear of Biolox®Delta. *Acta Biomater.* **2012**, *8*, 2348–2359. [CrossRef] [PubMed]

© 2019 by the authors. Licensee MDPI, Basel, Switzerland. This article is an open access article distributed under the terms and conditions of the Creative Commons Attribution (CC BY) license (http://creativecommons.org/licenses/by/4.0/).

Article

Neural Network Modelling of Track Profile in Cold Spray Additive Manufacturing

Daiki Ikeuchi [1,2,*], Alejandro Vargas-Uscategui [2], Xiaofeng Wu [1] and Peter C. King [2]

1. School of Aerospace, Mechanical and Mechatronic Engineering, The University of Sydney, Sydney, NSW 2006, Australia
2. Commonwealth Scientific and Industrial Research Organisation Manufacturing, Private Bag 10, Clayton, VIC 3169, Australia
* Correspondence: daiki.ikeuchi@sydney.edu.au

Received: 12 August 2019; Accepted: 29 August 2019; Published: 2 September 2019

Abstract: Cold spray additive manufacturing is an emerging technology that offers the ability to deposit oxygen-sensitive materials and to manufacture large components in the solid state. For further development of the technology, the geometric control of cold sprayed components is fundamental but not yet fully matured. This study presents a neural network predictive modelling of a single-track profile in cold spray additive manufacturing to address the problem. In contrast to previous studies focusing only on key geometric feature predictions, the neural network model was employed to demonstrate its capability of predicting complete track profiles at both normal and off-normal spray angles, resulting in a mean absolute error of 8.3%. We also compared the track profile modelling results against the previously proposed Gaussian model and showed that the neural network model provided comparable predictive accuracy, even outperforming in the predictions at cold spray profile edges. The results indicate that a neural network modelling approach is well suited to cold spray profile prediction and may be used to improve geometric control during additive manufacturing with an appropriate process planning algorithm.

Keywords: cold spray; neural network; additive manufacturing; model; spray angle; profile

1. Introduction

Cold spray is a materials deposition technology that is suitable for coatings and repair and is widely employed in industrial applications. This technology adopts a supersonic gas jet to accelerate powder particles to 500–1000 m/s and enables solid-state deposition onto a substrate by kinetic energy of the particles without melting. This mechanism offers unique characteristics that are difficult to achieve otherwise, including: Low oxygen-content deposition, the avoidance of melting-induced microstructure changes, and the ability to handle oxygen-sensitive materials without a protective atmosphere [1–3]. Furthermore, a high deposition rate can be achieved with a narrow nozzle diameter, resulting in a well-defined and high-density particle beam at small standoff distances [4].

The characteristics of cold spray are now recognized to offer great potential as an alternative solution to the field of additive manufacturing, namely Cold Spray Additive Manufacturing (CSAM) [5–9]. The elimination of a protective atmosphere environment provides the ability to fabricate larger manufactured components that are not possible with other additive manufacturing technologies, e.g., powder-bed additive manufacturing, while still allowing for excellent flexibility in the selection of oxygen-sensitive powder materials [9–11]. This benefit of cold spray technology can be further enhanced by the inclusion of a robotic system in CSAM which also allows the stability of fabrication, more Degrees of Freedom (DoF) for complex shapes and industrial automation [12–14]. Such robotic CSAM effectively utilizes its high deposition rate to produce components at industrially relevant part turnaround times [7,9]. Owing to the benefits of CSAM, successful demonstrations have been reported largely in aerospace industries at different levels of fabrication complexity: Simple rotational structures [15,16] and more complex components (e.g., fin arrays) [17–19].

However, the CSAM technology has not yet reached a mature technology level where it is considered a viable and reliable replacement to what is currently in use in commercial manufacturing industries due to a number of fundamental and practical problems. Fundamental problems are associated with the acceptable range of CSAM materials selection [20,21] and the microstructure and mechanical properties of deposits under different process parameters [22,23]. In contrast, practical challenges attract less attention from the CSAM community although providing a solution to them is a key aspect to facilitating the development of a commercial CSAM technology. One such practical challenge is the geometric control of as-fabricated components often associated with the nature of high production rate additive manufacturing technologies: namely, CSAM [8,9,24], Wire and Arc Additive Manufacturing (WAAM) [13,25] and Laser Cladding (LC) [26,27]. Low geometric control is attributed to a range of key issues that limit the application of additive manufacturing technologies such as the necessity of post-machining, difficulty in fabricating complex shapes, geometry-induced property variations and inconsistent quality of fabricated parts [8,9,28]. Therefore, addressing the challenge of geometric control is undoubtedly of great importance in CSAM as well as other high-speed additive manufacturing technologies.

From the perspective of geometric control, the development of a high-accuracy process model on the smallest processing unit (e.g., single cold spray track) offers a promising solution to the aforementioned problem since an aggregate of single tracks determines final part geometry. Furthermore, such a single-track model often plays a key role in the modelling of higher processing unit (i.e., overlapping and overlayer models) in the literature [25,29]. Previous studies of the single-track modelling fell into two main approaches: mathematical and data-driven modelling.

Suryakumar et al. approximated the profile of a single symmetric bead as a parabolic model in WAAM [30]. The model was developed in terms of WAAM process parameters as well as bead geometric characteristics (i.e., height and width). A second-order regression model was established with the aid of experiments to express the bead height in terms of the process parameters from which the bead width was calculated mathematically. This hybrid modelling approach showed reasonable pictorial agreement with a verification bead profile under the reported experimental conditions. Cai et al. employed a Gaussian model with a constant scaling coefficient to approximate the profile of a single symmetric cold spray track under different standoff distance scenarios [31]. The authors integrated the derived model into their Thermal Spray Toolkit, a software package in ABB RobotStudio®, for offline programming to predict cold spray track profiles.

Alternatively, a data-driven modelling approach attracted attention as an alternative to mathematical modelling approach with the increased accessibility of available software options. Mahapatra and Li applied an Artificial Neural Network (ANN) modelling with back propagation algorithm to predict the cross-sectional geometry of a single symmetric track profile in highly nonlinear and multivariate nature Pulsed-Laser Powder Deposition process [32]. The trained ANN model predicted bead width, cross-sectional area and heights at three segments within mostly 10% mean absolute error. Xiong et al. highlighted the development of ANN and second-order regression models in single symmetric bead geometry prediction in WAAM [33]. The authors compared the performance of the developed models in bead height and width predictions and reported that the ANN model outperformed in both predictions due to its ability to approximate any nonlinear process.

Despite the great capability of ANN modelling as seen in other additive manufacturing processes, it has drawn only a small amount of interest as a track modelling approach from the CSAM community. Furthermore, the application of the ANN modelling in prediction was greatly limited to key geometric characteristics only, e.g., height and width, in additive manufacturing [27,33]; such observations formed an underlying motivation to study in mathematical modelling that could describe more detailed geometric track profiles. This trend can be seen in previous CSAM studies focusing on the mathematical approach only (i.e., Gaussian model) to predict a single-track profile at both normal and off-normal spray angles [24,34]. However, a data-modelling approach can be more competent than what was previously conceived of in additive manufacturing as recently demonstrated successfully by Kochar et al. in joining application [35]. The data-driven approach offered great nonlinear mapping capability and multi-output predictions with affordable model complexity; such advantages are particularly desirable as asymmetric track profiles resulting from off-normal spray angles have become more frequent due to the necessity of complex spray strategies in CSAM. An accurate modelling of both symmetric and asymmetric single-track profiles with high geometric details will contribute to the improved geometric control in CSAM, enabling the fabrication of more complex and consistent parts with minimal post-machining.

In this study, we focus on the modelling of a single-track profile with high morphology in CSAM, both at normal and off-normal spray angles, using an ANN modelling to demonstrate its potential as a predictive modelling approach in additive manufacturing. The significance of this study is three-fold: (1) The application of a data-driven modelling approach in the prediction of a track profile to CSAM, (2) the modelling of an asymmetric track profile using the ANN model instead of the previous mathematical approach, and (3) the ANN modelling of a detailed track profile rather than key geometric characteristics only.

2. Materials and Methods

An ANN is a type of data-driven model for supervised machine learning which is sufficiently capable of handling nonlinearity and constructing an input–output relationship mapping based on a set of training data. The development of an effective ANN relies on a number of key design aspects such as input variable selection, data quality and network architecture [36,37]. In this study, three process variables were chosen as the inputs of the ANN model: spray angle, traverse speed, and standoff distance. These process variables are precisely controllable in real time with the support of an appropriate robotic system [12] and have been shown to be influential on cold spray geometric profiles in previous studies [24,31].

In this study, a full factorial approach was adopted to define the values of the input variables in the ANN training dataset due to the nonlinear and complex nature of CSAM and the affordable number of the input variables. In this approach, three levels were defined for traverse speed and standoff distance, while four levels were employed to capture the effects of spray angle on track profiles more precisely. The values of the input variables at each level are listed in Table 1. The lowest- and highest-level values were determined as those of the corresponding operating limits to maintain the sufficient deposit quality in the CSAM system. Defining the parameter boundaries at these operating limits avoided the weakness of an ANN model in extrapolation outside the training dataset [38]. The input values of intermediate levels were equally spaced between the lowest and highest level such that possible interactions between the input variables were adequately captured [39]. The resulting experiment design matrix required the fabrication of 36 samples for the ANN training dataset. The details of the experiment design matrix of each sample can be found in the Supplementary Materials (i.e., Tables S1 and S2).

Table 1. The levels of input variables in the experimental design matrix for the Artificial Neural Network (ANN) training dataset: 4 levels for spray angles, 3 levels for traverse speed and 3 levels for standoff distance.

Level	Spray Angle (°)	Traverse Speed (mm/s)	Standoff Distance (mm)
1	45	25	30
2	60	100	40
3	75	200	50
4	90	-	-

2.1. Sample Preparation

All sample preparations were performed using a commercial Impact Innovations (Haun, Germany) 5/11 cold spray gun guided by an ABB (Zurich, Switzerland) 4600 6-DoF robot. The cold spray gun was equipped with a long pre-chamber and an Impact Innovation's OUT1 tungsten carbide de Laval nozzle with a 6.2 mm exit diameter. Commercial purity grade-2 titanium from AP&C (Boisbriand, Canada) was selected as the powder feedstock. The particles were prepared by gas atomization and distributed within the size of 15 to 45 µm: (i.e., D_{10} = 19 µm, D_{50} = 34 µm and D_{90} = 45 µm). Nitrogen gas was preheated to 600 °C at a pressure of 5 MPa to accelerate the particles that were fed into the upstream of the nozzle at a feed rate of 1.9 kg/h. These spray variables except those listed in Table 1 were held constant throughout the sample fabrications. The substrate was a strip of commercial purity grade-2 titanium with a dimension of 6 × 30 × 200 mm. The surface of the substrate was prepared with a milling machine from Avemax Machinery (Taichung City, Taiwan) followed by grinding with P120-SiC emery paper from LECO (Moenchengladbach, Germany). Ethanol was used to clean the surface prior to the sample fabrications. The fabrication of all samples was randomized to obtain statistically unbiased results and minimize the effects of potential extraneous factors [40]. RobotStudio® software version 6.08 (ABB Robotics, Zurich, Switzerland) was used to verify that there was sufficient travel past the edge of the substrate to allow for the robot trajectory and traverse speed to stabilize prior to sample fabrications.

The profile of each sample was measured five times at randomly selected locations using a LEXT OLS4100 confocal laser scanning microscope from Olympus (Tokyo, Japan) and scanControl 2950-100 laser scanner from Micro-Epsilon (Ortenburg, Germany) with the z-axis measuring precision of at least 12 µm. The obtained measurements were processed with the in-built filtering: Flat Surface filtering in LEXT OLS4000 and average filtering with a filter size of 7 in scanControl Configuration Tools 6.0. The filtered profiles were averaged for each sample, resulting in the ANN output profiles considered in this study.

2.2. Artificial Neural Network Model Design and Training

In this study, a static multilayer perceptron ANN model was considered due to various successful demonstrations of its application as a predictive model in manufacturing processes. The model consisted of three different layer types: input layer, hidden layer, and output layer. Each layer contained a number of neurons with connections in between through activation functions. The number of neurons in the input layer corresponded to the number of input variables considered, i.e., 3 neurons in this study. The neurons in the hidden layer act as computational elements processing nonlinear mapping between the input and output variables and largely influence the performance and reliability of an ANN model [41]. Although a higher number of hidden neurons allow more accurate predictions or the modelling of more complex processes, it poses a higher risk of overfitting that is critical with only 36 training samples. For developing a reliable ANN model, this study iteratively investigated the performance of the ANN model with the different number of hidden neurons (i.e., 1 to 15 neurons) for each hidden layer. Similarly, the number of hidden layers was incrementally changed between 1 to 3 layers to optimize the ANN model architecture. Furthermore, the number of output neurons must be sufficient to achieve the objective of modelling a detailed track profile in CSAM. We adopted the area validation methodology proposed by Kochar et al. [35] in which polar lengths were considered as output neurons, measured from the tool center point as the origin. The number of output neurons was incrementally changed from 5 neurons with equal angular spacing between each output neuron (e.g., 45° each between 5 neurons) until the sufficient number was reached. Here, the sufficient number of output neurons was defined as that whose enclosed area reached and maintained at least 99% of the sample cross-sectional area in the last five consecutive candidates. An activation function is another critical aspect in an ANN model that computes the output of a neuron given the set of weights and biases as inputs. In this study, the commonly used hyperbolic tangent sigmoid and linear activation functions were chosen for hidden and output layers respectively. With the selected activation functions, all inputs and outputs variables were scaled to [−1 1] for improved training process [41].

The back-propagation algorithm with Bayesian regularization was selected as the training function of the ANN model. This algorithm depends on Levenberg-Marquardt optimization for updating weights and biases. The benefits of this training algorithm are two-fold: Robustness and the elimination of validation dataset, reducing the number of samples required [42]. In addition, the performance of the training process was measured using Mean Squared Error (MSE). The training of the ANN model was conducted using Deep Learning Toolbox in MATLAB® version R2018a and the training dataset in the supplementary material (i.e., Samples 1 to 36 in Table S1). To avoid the effects of different initial weights and biases, each ANN candidate model was retrained 100 times.

The performance of the trained ANN model was evaluated using an independent set of testing samples (see Table S2 in the Supplementary Material). The number of testing samples was determined according to the 75-25 training-testing data division method [43], resulting in a total of 12 testing samples (i.e., Samples 37-48). The values of the input variables in the testing dataset were randomly selected between their minimum and maximum operating limits with the aid of MATLAB® version R2018a.

3. Results and Discussion

3.1. Single-Track Profiles Validation

The quality of the cold spray profile samples was validated against the previous CSAM studies in terms of the effects of the input variables on the sample profiles. Figure 1 shows the effects of the following input variables on the profiles of the selected training samples: (a) Spray angle at 25 mm/s traverse speed and 30 mm standoff distance, (b) Traverse speed at 90° spray angle and 30 mm standoff distance, and (c) Standoff distance at 90° spray angle and 25 mm/s traverse speed.

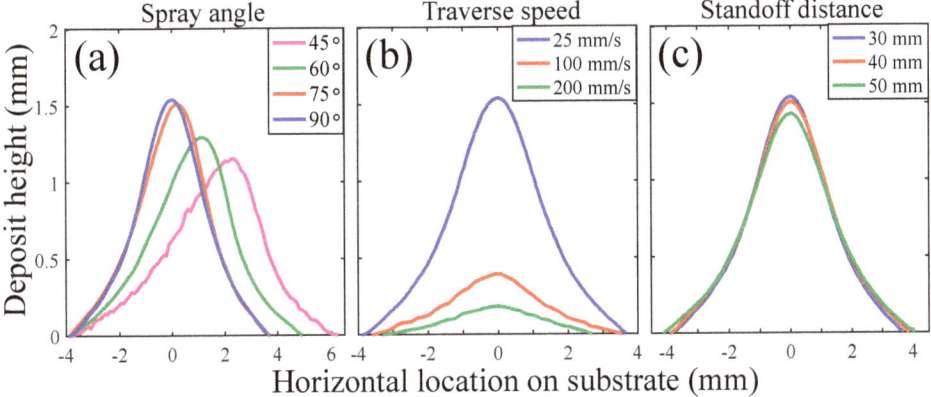

Figure 1. (a) the effect of spray angle at 25 mm traverse speed and 30 mm standoff distance (45°—sample 1, 60°—sample 2, 75°—sample 3, 90°—sample 4), (b) the effect of traverse speed at 90° spray angle and 30 mm standoff distance (25 mm/s—sample 4, 100 mm/s—sample 8, 200 mm/s—sample 12) and (c) the effect of standoff distance at 90° spray angle and 25 mm/s traverse speed (30 mm—sample 4, 40 mm—sample 16, 50 mm—sample 28).

In Figure 1a, it is clear that the spray angle was positively correlated to the height and negatively to the width of the sample profiles, being consistent with the previous studies [24,44]. The smaller effect of spray angle between 75° and 90° was attributed to the smaller relative deposition efficiency drop; in comparison, such phenomenon was observed between 80° and 90° spray angle in [24].

Importantly, the effect of traverse speed was found to be nonlinear and the most influential on the track profiles in Figure 1b. This observation suggests that the more levels of traverse speed may be employed in the experimental design matrix for the ANN training dataset to integrate more relevant information into an ANN model, especially at low traverse speeds (i.e., between 25 and 100 mm/s). The lower traverse speed resulted in a thicker and sharper track profile, while widening the track profile as also seen in [9,24].

In contrast, the effect of standoff distance was the least influential on the track profiles in Figure 1c. With the larger standoff distance, the track profile became shorter and wider as observed in [31]. This phenomenon was previously confirmed by Pattison et al. [45] and indicated that the standoff distance parameter space covered in this study was in the medium region near the optimal deposition efficiency point. It is of great interest to study further towards the nonlinear extreme ends (e.g., 10 and 100 mm), maximizing the benefits of nonlinear mapping ability in an ANN model.

In summary, the validation of the selected track profiles confirmed that the fabricated profiles were consistent with the previously reported trends and therefore of sufficient quality as the output profile data considered in this study. Note that the track profiles of all samples are presented in Figures S1–S3 in the Supplementary Materials.

3.2. Neural Network Architecture Validation

The area validation method for determining the sufficient number of output neurons was performed over all the samples and the results of some randomly selected samples are shown in Figure 2 to illustrate the trend of area convergence. The mean sufficient number of output neurons was found as 67 over all the samples, while the maximum number was 167. Here, 67 output neurons were chosen for the ANN model, taking polar lengths from the tool center point at every 2.72°. This selection was because the maximum sufficient number of output neurons resulted in capturing too fine geometric features that could be considered as noises. Furthermore, the fewer number of output neurons allows a simpler ANN architecture, thereby reducing the computational burden of training process, while the resulting ANN model is accurate enough to achieve the objective of describing a detailed CSAM track profile. The resulting output neuron parameters are presented in Tables S3 and S4 in the Supplementary Materials.

The iterative investigation of different ANN hidden layer topologies concluded that two hidden layers with 6 and 10 neurons achieved the best predictive performance on the normalized independent testing dataset with MSE of 0.009454 and R^2 coefficient of 0.9493 (see Figure 3). The mean predictive performance for each output geometric point was evaluated among all 12 testing samples and the corresponding overall predictive performance for all 67 outputs is summarized in Table 2. Both Mean Absolute Percent Error (MAPE) and Maximum Absolute Percent Error (MXAPE) were reasonable in comparison with the previous studies in different manufacturing processes (i.e., MAPE = just below 10%, MXAPE ≈ 11% [35] and MAPE = 6.611%, MXAPE = 10.31% [46]). Consequently, the results demonstrated the suitability of a data-driven ANN modelling approach to the prediction of a track profile in CSAM.

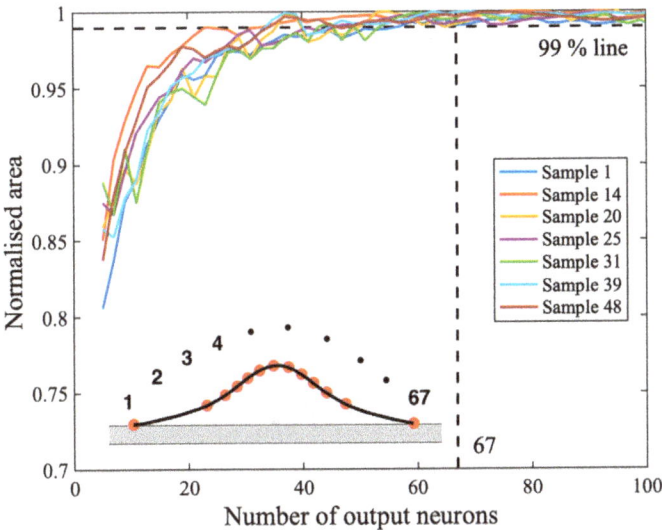

Figure 2. The validation results of the randomly selected samples. The validation of the number of output neurons indicates that the mean 67 output neurons are sufficient to describe the track profile with high fidelity.

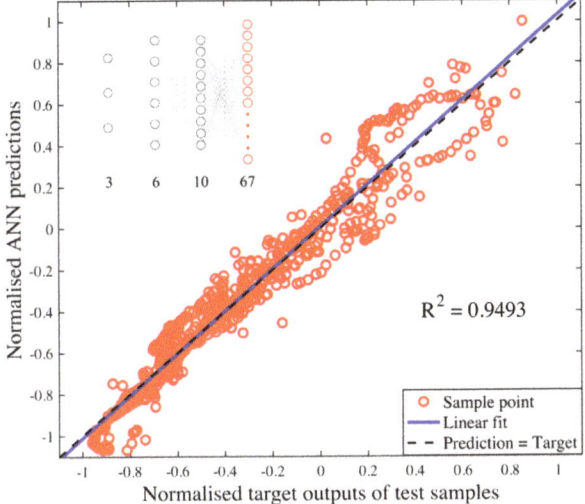

Figure 3. Normalized ANN predictions vs. target output neuron values of all test samples. The ANN had the architecture of [3 6 10 67], resulting in MSE of 0.009454 and $R^2 = 0.9493$.

Table 2. Summary of the performance evaluation results of the developed model in Figure 3 in terms of Mean Absolute Error (MAE), Maximum Absolute Error (MXAE), Mean Absolute Percent Error (MAPE), and Maximum Absolute Percent Error (MXAPE).

MAE (mm)	MXAE (mm)	MAPE (%)	MXAPE (%)
0.05782	0.1522	8.342	10.20

3.3. Evaluation of Artificial Neural Network Modelling for Predicting Single-Track Profiles

Figure 4 shows the track profile of the two selected test samples as an illustration: (a) Sample 37 at a nearly normal spray angle (i.e., 86°) and (b) Sample 39 at a spray angle of 48°. The developed ANN model was used to predict the track profiles, resulting in a qualitatively good agreement with the measured profiles. The MAEs were 0.009550 mm and 0.04256 mm for Sample 37 and 39, respectively. Thus, it is demonstrated that the application of an ANN modelling approach is possible to predict both symmetric and asymmetric track profiles at normal and off-normal spray angles. However, for Sample 39 at a lower spray angle, a larger deviation from the measured profile was found, as compared to Sample 37, in the high region of the track profile (between 3 and 7 mm on the substrate). The possible causes for this observation include: (1) The lack of training samples within this region to provide sufficient robustness to external factors (e.g., robot joint misalignment and tool centre point variation) at an off-normal spray angle and (2) inefficient experimental design matrix to capture high nonlinearities in CSAM, e.g., more traverse speed levels may be suitable as discussed in Section 3.1 towards the low-speed end, resulting in thicker track profiles. The robustness issue was also raised in the application of ANN modelling in welding [35], but it is more severe when a large number of ANN output predictions is necessary with a small number of input parameters such as in this study. Note that the ANN prediction results for all other test samples are presented in Table S5 and graphically shown in Figure S4 in the Supplementary Materials.

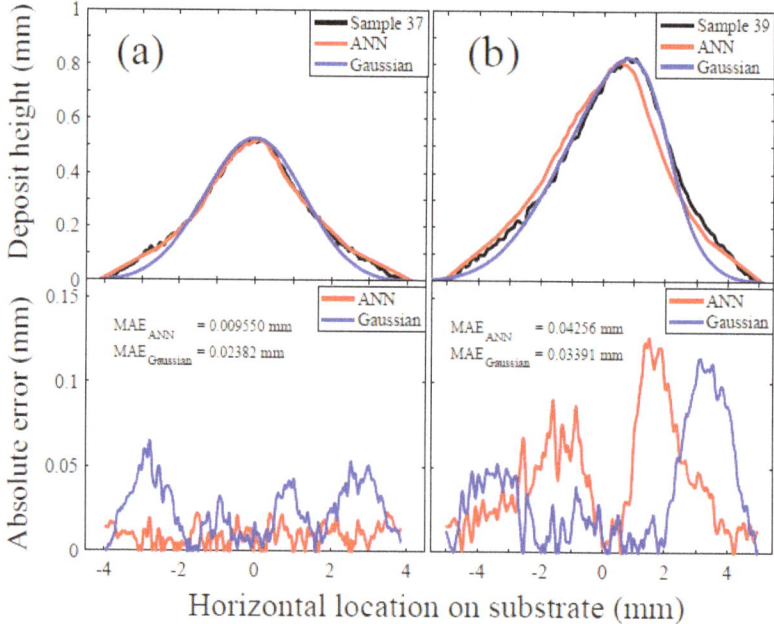

Figure 4. The experimental single-track profiles of the two selected samples are shown as illustrative examples along with the corresponding ANN (red) and Gaussian (blue) models: (**a**) Sample 37 (spray angle: 86°, traverse speed: 75 mm/s, standoff distance: 45 mm) and (**b**) Sample 39 (spray angle: 48°, traverse speed: 34 mm/s, standoff distance: 41 mm).

To demonstrate the potential of ANN modelling to predict detailed track profile for the objective of this study, in Figure 4, we also compared the ANN modelling results against the Gaussian modelling approach in cold spray proposed by Chen et al. [24]. The details of the Gaussian models can be found in Table S6 in the Supplementary Material. The ANN modelling approach showed about 2.5 times smaller MAE than the Gaussian model for Sample 37, but about 1.3 times larger MAE for Sample 39. The latter result was mainly attributed to the larger deviation at the high portion of the track profile as discussed previously. Meanwhile, the ANN modelling showed better predictive performance in the region of track profile edges than the Gaussian model. Such better predictive performance at profile edges was most likely due to the ANN model adequately capturing the complex multivariate nature of cold spray process (e.g., bow shock and compressed gas layer [47]), while this was observed to be lacking with the Gaussian model used previously in cold spray [24,31]. In summary, the comparative study of the two modelling approaches in Figure 4 showed that the ANN modelling possessed the potential to provide the prediction of detailed track profiles in CSAM at the same level of accuracy or higher.

4. Conclusions

This study demonstrated the potential of a data-driven modelling approach in the prediction of single-track profiles in CSAM, rather than only key geometric features as in previous studies. The ANN modelling enabled an accurate description of track profiles at even off-normal spray angles that are frequently encountered during the cold spray process of complex shapes. Furthermore, the detailed track profiles predicted by the ANN model were in good qualitative agreement with the measured profiles, even outperforming at the region of profile edges as compared to the previously proposed Gaussian modelling approach. Therefore, the data-driven modelling, in combination with an appropriate process planning algorithm, possesses the potential to improve the problem of geometric control in additive manufacturing processes and therefore foster the development of a commercial CSAM technology. With the appropriate adjustment of ANN input feature parameters and architecture, the approach presented in this study can be extended to other additive manufacturing techniques such as WAAM and LC.

However, the limitation of the ANN modelling approach was also encountered due to the size of training dataset and robustness. These issues were more significant in this study as the ANN approach adopted a larger number of output neurons than previous studies where only key geometric features were predicted. Therefore, it is of great importance in future works that a more data-efficient modelling approach is explored, and real-time measurement and a data processing system are developed so that the data diversity and collection rate increase. Furthermore, the comparative study of the two models showed that the Gaussian model predicted with better accuracy within the high portion of track profiles, while the ANN model was more accurate towards the profile edges. This finding triggers a motivation for exploring a hybrid modelling approach in future works, taking advantages from the two modelling approaches, while minimizing the disadvantages discussed in this study.

In addition to the aforementioned future works, we plan to extend this study to overlapping and overlayer modellings and integrate the ANN model from this study into our toolpath planning algorithm at a system level.

Supplementary Materials: The following are available online at http://www.mdpi.com/1996-1944/12/17/2827/s1, Figure S1: the measured track profiles of Sample 1 to 18, Figure S2: the measured track profiles of Sample 19 to 36, Figure S3: the measured track profiles of Sample 37 to 48, Figure S4: the predicted track profiles using the developed ANN model, Table S1: Input parameters in the training dataset, Table S2: Input parameters in the testing dataset, Table S3: Output parameters in the training dataset, Table S4: Output parameters in the testing dataset, Table S5: ANN results for all test samples, Table S6: Gaussian model parameters.

Author Contributions: Conceptualization, D.I.; Methodology, D.I., A.V.U., P.K.; Software, D.I.; Validation, D.I.; Formal analysis, D.I.; Investigation, D.I.; Data curation, D.I.; Writing—original draft preparation, D.I.; Writing—review and editing, D.I., A.V.U., X.W., P.K.; Supervision, X.W., P.K.

Funding: This research was funded by CSIRO's Active Integrated Matter Future Science Platform (AIM FSP) under the testbed number: TB10_WB04.

Acknowledgments: The authors would like to acknowledge the Sydney Informatics Hub and the University of Sydney's high-performance computing cluster, Artemis, for providing the high-performance computing resources that contributed to the results presented in this paper.

Conflicts of Interest: The authors declare no conflicts of interest.

References

1. Gärtner, F.; Stoltenhoff, T.; Schmidt, T.; Kreye, H. The cold spray process and its potential for industrial applications. *J. Therm. Spray Techn.* **2006**, *15*, 223–232. [CrossRef]
2. Villafuerte, J. Current and future applications of cold spray technology. *Met. Finish.* **2010**, *108*, 37–39. [CrossRef]
3. Luo, X.-T.; Li, C.-X.; Shang, F.-L.; Yang, G.-J.; Wang, Y.-Y.; Li, C.-J. High velocity impact induced microstructure evolution during deposition of cold spray coatings: A review. *Surf. Coat. Tech.* **2014**, *254*, 11–20. [CrossRef]
4. Karthikeyan, J. The advantages and disadvantages of the cold spray coating process. In *The Cold Spray Materials Deposition Process*; Champagne, V.K., Ed.; Woodhead Publishing: New York, NY, USA, 2007; pp. 62–71.
5. Pattison, J.; Celotto, S.; Morgan, R.; Bray, M.; O'Neill, W. Cold gas dynamic manufacturing: A non-thermal approach to freeform fabrication. *Int. J. Mach. Tool. Manu.* **2007**, *47*, 627–634. [CrossRef]
6. Sova, A.; Grigoriev, S.; Okunkova, A.; Smurov, I. Potential of cold gas dynamic spray as additive manufacturing technology. *Int. J. Adv. Manuf. Techn.* **2013**, *69*, 2269–2278. [CrossRef]
7. Pathak, S.; Saha, G. Development of sustainable cold spray coatings and 3D additive manufacturing components for repair/manufacturing applications: A critical review. *Coatings* **2017**, *7*, 122. [CrossRef]
8. Li, W.; Yang, K.; Yin, S.; Yang, X.; Xu, Y.; Lupoi, R. Solid-state additive manufacturing and repairing by cold spraying: A review. *J. Mater. Sci. Technol.* **2018**, *34*, 440–457. [CrossRef]
9. Yin, S.; Cavaliere, P.; Aldwell, B.; Jenkins, R.; Liao, H.; Li, W.; Lupoi, R. Cold spray additive manufacturing and repair: Fundamentals and applications. *Addit. Manuf.* **2018**, *21*, 628–650. [CrossRef]
10. Mutombo, K. Research and Development of Ti and Ti alloys: Past, present and future. *IOP Conf. Ser: Mater. Sci. Eng.* **2018**, *430*, 0120071–0120076. [CrossRef]
11. Titomic Titomic Kinetic Fusion™. Available online: https://www.titomic.com/titomic-kinetic-fusion.html (accessed on 22 March 2019).
12. Danielsen Evjemo, L.; Moe, S.; Gravdahl, J.T.; Roulet-Dubonnet, O.; Gellein, L.T.; Brøtan, V. Additive Manufacturing by Robot Manipulator: An Overview of The State-of-The-Art and Proof-of-Concept Results. In Proceedings of the 22nd IEEE International Conference on Emerging Technologies and Factory Automation (ETFA), Limassol, Cyprus, 12–15 September 2017; pp. 1–8.
13. Ma, G.; Zhao, G.; Li, Z.; Yang, M.; Xiao, W. Optimization strategies for robotic additive and subtractive manufacturing of large and high thin-walled aluminum structures. *Int. J. Adv. Manuf. Techn.* **2019**, *101*, 1275–1292. [CrossRef]
14. Zhang, Y.J.; Li, W.B.; Li, D.Y.; Xiao, J.K.; Zhang, C. Modeling of thickness and profile uniformity of thermally sprayed coatings deposited on cylinders. *J. Therm. Spray Techn.* **2018**, *27*, 288–295.
15. Barnett, B.; Trexler, M.; Champagne, V. Cold sprayed refractory metals for chrome reduction in gun barrel liners. *Int. J. Refract. Met. H.* **2015**, *53*, 139–143. [CrossRef]
16. King, P.; Gulizia, S.; Urban, A.; Barnes, J. Process for producing A Preform using Cold Spray. Available online: https://patents.google.com/patent/US20170157671A1/en (accessed on 4 August 2019).
17. Lynch, M.E.; Gu, W.; El-Wardany, T.; Hsu, A.; Viens, D.; Nardi, A.; Klecka, M. Design and topology/shape structural optimisation for additively manufactured cold sprayed components. *Virtual Phys. Prototy.* **2013**, *8*, 213–231. [CrossRef]
18. Cormier, Y.; Dupuis, P.; Jodoin, B.; Corbeil, A. Pyramidal fin arrays performance using streamwise anisotropic materials by cold spray additive manufacturing. *J. Therm. Spray Techn.* **2016**, *25*, 170–182. [CrossRef]
19. Perry, J.; Richer, P.; Jodoin, B.; Matte, E. Pin fin array heat sinks by cold spray additive manufacturing: Economics of powder recycling. *J. Therm. Spray Techn.* **2019**, *28*, 144–160. [CrossRef]
20. AL-Mangour, B.; Dallala, R.; Zhim, F.; Mongrain, R.; Yue, S. Fatigue behavior of annealed cold-sprayed 316L stainless steel coating for biomedical applications. *Mater. Lett.* **2013**, *91*, 352–355. [CrossRef]
21. Poza, P.; Múnez, C.J.; Garrido-Maneiro, M.A.; Vezzù, S.; Rech, S.; Trentin, A. Mechanical properties of Inconel 625 cold-sprayed coatings after laser remelting. Depth sensing indentation analysis. *Surf. Coat. Tech.* **2014**, *243*, 51–57. [CrossRef]
22. Gärtner, F.; Stoltenhoff, T.; Voyer, J.; Kreye, H.; Riekehr, S.; Koçak, M. Mechanical properties of cold-sprayed and thermally sprayed copper coatings. *Surf. Coat. Tech.* **2006**, *200*, 6770–6782. [CrossRef]
23. Suhonen, T.; Varis, T.; Dosta, S.; Torrell, M.; Guilemany, J.M. Residual stress development in cold sprayed Al, Cu and Ti coatings. *Acta Mater.* **2013**, *61*, 6329–6337. [CrossRef]

24. Chen, C.; Xie, Y.; Verdy, C.; Liao, H.; Deng, S. Modelling of coating thickness distribution and its application in offline programming software. *Surf. Coat. Techn.* **2017**, *318*, 315–325. [CrossRef]
25. Ding, D.; Pan, Z.; Cuiuri, D.; Li, H. A multi-bead overlapping model for robotic wire and arc additive manufacturing (WAAM). *Robot. Cim-Int. Manuf.* **2015**, *31*, 101–110. [CrossRef]
26. Saqib, S.; Urbanic, R.J.; Aggarwal, K. Analysis of laser cladding bead morphology for developing additive manufacturing travel paths. *Procedia CIRP* **2014**, *17*, 824–829. [CrossRef]
27. Liu, H.; Qin, X.; Huang, S.; Jin, L.; Wang, Y.; Lei, K. Geometry characteristics prediction of single track cladding deposited by high power diode laser based on genetic algorithm and neural network. *Int. J. Precis. Eng. Man.* **2018**, *19*, 1061–1070. [CrossRef]
28. Frazier, W.E. Metal additive manufacturing: A review. *J. Mater. Eng. Perform.* **2014**, *23*, 1917–1928. [CrossRef]
29. Nenadl, O.; Kuipers, W.; Koelewijn, N.; Ocelík, V.; De Hosson, J.T.M. A versatile model for the prediction of complex geometry in 3D direct laser deposition. *Surf. Coat. Tech.* **2016**, *307*, 292–300. [CrossRef]
30. Suryakumar, S.; Karunakaran, K.P.; Bernard, A.; Chandrasekhar, U.; Raghavender, N.; Sharma, D. Weld bead modeling and process optimization in Hybrid Layered Manufacturing. *Comput. Des.* **2011**, *43*, 331–344. [CrossRef]
31. Cai, Z.; Deng, S.; Liao, H.; Zeng, C.; Montavon, G. The Effect of Spray Distance and Scanning Step on the Coating Thickness Uniformity in Cold Spray Process. *J. Therm. Spray Techn.* **2014**, *23*, 354–362. [CrossRef]
32. Mahapatra, M.M.; Li, L. Prediction of pulsed-laser powder deposits' shape profiles using a back-propagation artificial neural network. *P. I. Mech. Eng. B-J. Eng.* **2008**, *222*, 1567–1576. [CrossRef]
33. Xiong, J.; Zhang, G.; Hu, J.; Wu, L. Bead geometry prediction for robotic GMAW-based rapid manufacturing through a neural network and a second-order regression analysis. *J. Intell. Manuf.* **2014**, *25*, 157–163. [CrossRef]
34. Klinkov, S.V.; Kosarev, V.F.; Ryashin, N.S.; Shikalov, V.S. Influence of particle impact angle on formation of profile of single coating track during cold spraying. In Proceedings of the AIP Conference Proceedings, Novosibirsk, Russia, 13–19 August 2018; Volum 2027, pp. 0200071–0200076.
35. Kochar, P.; Sharma, A.; Suga, T.; Tanaka, M. Prediction and control of asymmetric bead shape in laser-arc hybrid fillet-lap joints in sheet metal welds. *Lasers Manuf. Mater. Process.* **2019**, *6*, 67–84. [CrossRef]
36. Shafi, I.; Ahmad, J.; Shah, S.I.; Kashif, F.M. Impact of Varying Neurons and Hidden Layers in Neural Network Architecture for a Time Frequency Application. In Proceedings of the 2006 IEEE International Multitopic Conference, Islamabad, Pakistan, 23–24 December 2006; pp. 188–193.
37. May, R.; Dandy, G.; Maier, H. Review of Input Variable Selection Methods for Artificial Neural Networks. In *Artificial Neural Networks—Methodological Advances and Biomedical Applications*; InTechOpen: London, UK, 2011; pp. 19–44.
38. Barnard, E.; Wessels, L. Extrapolation and interpolation in neural network classifiers. *IEEE Contr. Syst. Mag.* **1992**, *12*, 50–53.
39. Noriega, A.; Blanco, D.; Alvarez, B.J.; Garcia, A. Dimensional accuracy improvement of FDM square cross-section parts using artificial neural networks and an optimization algorithm. *Int. J. Adv. Manuf. Technol.* **2013**, *69*, 2301–2313. [CrossRef]
40. Krishaniah, K.; Shahabudeen, P. Fundamentals of experimental design. In *Applied Design of Experiments and Taguchi Methods*; PHI Learning Pvt. Ltd.: New Delhi, India, 2012; pp. 22–48.
41. Haykin, S. *Neural Networks and Learning Machines*, 3rd ed.; Pearson Education, Inc.: New Jersey, NJ, USA, 2009.
42. Burden, F.; Winkler, D. Bayesian regularization of neural networks. *Method. Mol. Biol.* **2008**, *458*, 25–44.
43. Alibakshi, A. Strategies to develop robust neural network models: Prediction of flash point as a case study. *Anal. Chim. Acta* **2018**, *1026*, 69–76. [CrossRef] [PubMed]
44. Luo, X.-T.; Li, Y.-J.; Li, C.-X.; Yang, G.-J.; Li, C.-J. Effect of spray conditions on deposition behavior and microstructure of cold sprayed Ni coatings sprayed with a porous electrolytic Ni powder. *Surf. Coat. Tech.* **2016**, *289*, 85–93. [CrossRef]
45. Pattison, J.; Celotto, S.; Khan, A.; O'Neill, W. Standoff distance and bow shock phenomena in the Cold Spray process. *Surf. Coat. Tech.* **2008**, *202*, 1443–1454. [CrossRef]

46. Katherasan, D.; Elias, J.V.; Sathiya, P.; Haq, A.N. Simulation and parameter optimization of flux cored arc welding using artificial neural network and particle swarm optimization algorithm. *J. Intell. Manuf.* **2014**, *25*, 67–76. [CrossRef]
47. Kotoban, D.; Grigoriev, S.; Okunkova, A.; Sova, A. Influence of a shape of single track on deposition efficiency of 316L stainless steel powder in cold spray. *Surf. Coat. Tech.* **2017**, *309*, 951–958. [CrossRef]

© 2019 by the authors. Licensee MDPI, Basel, Switzerland. This article is an open access article distributed under the terms and conditions of the Creative Commons Attribution (CC BY) license (http://creativecommons.org/licenses/by/4.0/).

MDPI
St. Alban-Anlage 66
4052 Basel
Switzerland
Tel. +41 61 683 77 34
Fax +41 61 302 89 18
www.mdpi.com

Materials Editorial Office
E-mail: materials@mdpi.com
www.mdpi.com/journal/materials

www.ingramcontent.com/pod-product-compliance
Lightning Source LLC
LaVergne TN
LVHW070701100526
838202LV00013B/1011